Unrewarded

The discovery of our universe in 42
Nobel Prizes that were never awarded

Ben Moore
Professor of Astrophysics
University of Zurich

Copyright © 2023 Ben Moore
All rights reserved for the English text.
Foreign language rights © Kein & Aber AG Zurich-Berlin
Published in German by Kein & Aber in 2022: "Sternenstaub"
www.benmoore.ch

Contents

Foreword 1
Part I: The first scientists 5
1. Eratosthenes c276-c194BC *"for demonstrating the Earth is a sphere and measuring its size"* 7
2. Aristarchus c310-c230BC *"for discovering our place in the solar system"* 12
3. Nicolaus Copernicus (1473-1543) *"for the heliocentric model of our solar system"* 20
4. Johannes Kepler (1571-1630) *"for the laws of planetary motion"* 25
5. Galileo Galilei (1564-1642) *"for the evidence that the planets orbited the Sun and for the discovery of our galaxy"* 31
6. Isaac Newton (1642-1726) *"for demonstrating that a force called gravity gives rise to elliptical planetary orbits"* 38
7. Edmund Halley (1656-1742) *"for shattering the crystalline spheres and for the technique for measuring the distance to the Sun"* 46
8. Immanuel Kant (1724-1804) *"for his visionary ideas of the origin of the solar system and of a vast and evolving universe"* 54
9. Henry Cavendish (1731-1810) *"for determining the density of the Earth and the absolute strength of gravity"*. 60
10. Friedrich Bessel (1784-1846) *"for measuring the distances to the stars"* 66

Part II: The scientists who discovered our place in time and space 71

11. Henri Poincaré (1854-1912) *"for the principle of relativity, the discovery of chaos theory and for predicting the existence of gravitational waves"* 73

12. Albert Einstein (1879-1955) *"for the theory of general relativity that connects space, time, matter and energy"* 82

13. Henrietta Swan Leavitt (1868-1921) *"for the discovery of an astronomical standard candle that led to the determination of our place in the Milky Way galaxy and within the universe"* 91

14. Georges Lemaître (1894-1966) *"for the discovery that the universe is expanding and had a hot and dense beginning – the big bang"* 98

15. Arthur Eddington (1882-1944) *"for the energy source and the first theoretical models of stars"* 105

16. Cecilia Payne (1900-1979) *"for discovering what stars are made of"* 115

17. George Gamow (1904-1968) *"for the theory of big bang nucleosynthesis"* 124

18. Ralph Alpher (1921-2007) & Robert Herman (1914-1997) *"for predicting the measurable afterglow of the big bang – the cosmic microwave background"* 130

19. Clair Patterson (1922-1995) *"for the measuring the age of the solar system"* 140

20. Allan Sandage (1926-2010) *"for measuring the age of the galaxy and the universe"* 147

21. Fred Hoyle (1915-2001) *"for the theory that elements are forged inside stars"* 154

22. Fritz Zwicky (1898-1974) *"for discovering dark matter in galaxy clusters, interpreting supernova and predicting the existence of neutron stars"* 164

23. Vera Rubin (1928 – 2016) *"for the observations that convinced scientists that dark matter exists"* 177

24. Rainer Sachs (1932-), Arthur Wolfe (1939-2014) and Joe Silk (1942-) *"for predicting temperature variations in the cosmic microwave background photons"* 186

25. Marietta Blau (1894-1970) *"for developing the photographic method of studying nuclear processes and for discovering cosmic ray disintegration of atoms"* 193

26. Lise Meitner (1878-1968) *"for the theory and discovery of nuclear fission"* 200

27. Yakov Zeldovich (1914-1987) *"for theoretical discoveries in physical cosmology"* 208

28. Fabiola Gianotti 1960- and 3000 others *"for recreating the conditions within the first second of the big bang and for discovering the Higgs boson"* 217

29. Jocelyn Bell Burnell (1943-) *"for the discovery of neutron stars"* 225

30. Shiv Kumar (1939-) *"for predicting the minimum mass of a star, the existence of brown dwarfs and population III stars"* 231

31. Aleksander Wolszczan (1946-) and Dale Frail (1961-) *"for the discovery of extra-solar planets"* 238

32. Carl Sagan (1934-1996) *"for determining the past and future of the Earth in the face of an evolving Sun, and for inspiring us all"* 245

33. Emmy Noether (1882-1935) *"for the connection between space and time and conservation laws"* 253

34. Chien-Shiung Wu (1912-1997) *"for showing that nature is not always symmetric"* 260

35. Edward Tryon (1940-2019) *"for the idea that the universe could begin from nothing via a quantum fluctuation"* 267

36. Ludwig Boltzmann (1844-1906) *"for the interpretation of thermodynamics and the arrow of time"* 274

37. Pier Giorgio Merli (1943-2008), Gian Franco Missiroli (1945-) & Giulio Pozzi (1945-) *"for the most beautiful experiment ever performed"* 283

38. Hugh Everett (1930-1982) *"for the many worlds interpretation of quantum mechanics"* 291

39. Stephen Hawking (1942-2018) *"for showing that even black holes will not last forever"* 300

40. Freeman Dyson (1923-2020) *"for contributions to quantum field theory and for his visionary work on the future of life in our universe"* 307

41. Erast Gliner (1923-2021) *"for the theory of inflation"* 315

42. Andrei Linde (1948-) & Paul Steinhardt (1952-) *"for eternal inflation and the theory of the multiverse"* ... 325

    Epilogue 334

    Appendix I: Misconceptions about the big bang: the tireless ant analogy to our expanding universe 338

    Appendix II: Spacetime 343

Foreword

Many of the scientists who made great discoveries in astronomy, astrophysics and cosmology were unrewarded for their efforts. The discoveries are often credited to the wrong scientist, or grand prizes given to the wrong person. More often than not, the fundamental contributions of others have been neglected or forgotten.

For example, if you ask a scientist, who discovered that the universe was expanding, that it had a beginning, you will most likely be told 'Edwin Hubble'. In fact, Hubble did not discover the big bang. Hubble never even believed in the idea that the universe was expanding. This phenomenon was actually discovered by a Belgian priest. You might even be less aware that the discovery of our place in the universe was thanks to Henrietta Leavitt, who was employed as a human 'computer' to catalogue the brightness of stars.

One of the greatest honours for a scientist is to have their discovery live on in their name. Another great honour is to receive the recognition of famous named prizes, and none are as famous as the Nobel Prize, the Oscar of science.

Alfred Nobel was a Swedish chemist, inventor and business man. In 1888 his brother died, and a French newspaper mistakenly published Alfred's obituary. It condemned him for his invention of military explosives and stated "*Le marchand de la mort est mort*". It went on to say that "Dr. Alfred Nobel, who became rich by finding ways to kill more people faster than ever before, died yesterday." This made Alfred Nobel, who was alive and well, reflect on how he wished to be remembered after death. In 1895 he signed his last will and testament, leaving nearly all his fortune to establish the five Nobel Prizes in physics, chemistry, medicine, literature and peace. The amount he left would be equivalent to about half a billion euros today.

Nobel Prize winners currently share about one million euros in each category every year. But the real prize is more than that since it is the greatest honour for any scientist to receive. There are prizes that give out more money, but they are not as well known or as prestigious. We hear a lot about those scientists who receive a Nobel Prize, but what about those who do not?

Between 1901 and 2023, there have been 224 Nobel Laureates in physics. For most of this time astrophysics and cosmology were ignored. Moreover, some of these prizes were incorrectly awarded and women have been notoriously overlooked. Of those 224 Nobel Prizes only five have been given to women. Although great progress has been made in equality, there is still a long way to go. Women were not allowed to undertake science for most of our scientific history. Yet even in the twentieth century there have been many women who were overlooked in the face of their male colleagues.

This is the story of the discovery of our universe with my personal view on who would have deserved the recognition of a Nobel Prize. I have carried out research in astrophysics and cosmology for over thirty years, but it is not an easy task. And probably my colleagues might question one or more of the people on my list. Dear colleagues, excuse my lack of mention of your favourite scientist hero, or even yourselves. And how should we decide on who deserves such an award? Is it the suggestion of the idea? Or those who turn that idea into a result through experiment or observation? What I have learned in my own research career is that a new idea is the most precious thing.

My story starts long ago in time. In his will, Alfred Nobel did not specify that his prizes should only be awarded to people who were still alive. That was only added to the Nobel Foundation statutes in 1974. Excuse my literal interpretation to give credit where credit is due as there are some brilliant scientists who died long before Mr. Nobel, yet without their insights into the cosmos we would not be where we are. After all, science proceeds step by step, with wrong paths and dead ends, together with serendipity luck and remarkable insights.

His will also states that the prize should be awarded to one person. The Nobel Foundation changed its statutes to award the prize to as many as three people for two different discoveries. That also seems a rather arbitrary modification and has led to many conflicts. This is particularly problematic in today's research environment where hundreds of scientists may be working on one experiment.

In Alfred Nobel's will he wrote how his assets should be disbursed as an annual series of prizes. It states *'The interest is to be divided into five equal parts and distributed as follows: one part to the person who made the most important discovery or invention in the field of physics;...'* This is interpreted by the Nobel Committee as a discovery or theory that has been confirmed. I find this interpretation unsatisfactory. Discoveries in science give support to theories but theories will always remain

3

theories and do not become facts. That life evolves is a theory, that the universe is expanding is a theory. At some point, any theory may be replaced by a better one, although it must also incorporate all the successes of the theory it replaces. That is why I believe that groundbreaking theories and ideas also deserve a Nobel Prize. These are profound insights that add to our understanding and interpretation of our universe and deserve recognition.

Why 42? Aside from the connection with my favourite science fiction book, when I first wrote down a list of the most important discoveries that had not received a Nobel Prize it totalled 42. One could argue there should be less, or even more. Of course, the answer to life, the universe and everything will require the work of many future brilliant scientists, but those described in this story have set us a long way towards this goal.

I will tell my ideas as a story, flowing between people and through time. After all, the discovery of our universe is the greatest story of all.

Ben Moore, Zurich

Part I:
The first scientists

How did we manage to figure out that the world could be comprehended? At what point did some people begin to discard mysticism, gods and spirits and start to explain natural phenomena as a consequence of cause and effect? Allow me to reward the work of ten philosophers and early scientists who set us on the path towards understanding our place in space and time, but who lived and died before the Nobel Prize was first awarded in 1901. Of course, this list could be far longer, but I think it was the discoveries of these ten individuals which were fundamental in leading us to the discovery of our universe in the 20th century.

The Greek islands and the shores of the Aegean Sea provided an ideal environment for abundant food, prosperity and technological advancement. Perhaps it was these circumstances that allowed its citizens time for intellectual and creative thought. But I often wonder why none of the great civilisations that came before, from the Harrapans to the Babylonians, ever attempted naturalistic interpretations of what they saw. For thousands of years the planets and stars were the homes of the gods, believed to be responsible for all natural phenomena from earthquakes to the weather, and indeed for most aspects of the world and people's lives. Then rather suddenly, a few individuals began to question the workings of the natural world – they tried to explain things through cause and effect, through observations and experiments, through thought and logic.

There is a long list of ancient Greeks who contributed to the quest for knowledge and understanding of our world and the cosmos. Democritus and his mentor Leucippus were the first to believe that all things are made of tiny particles so small that they cannot be seen. They called them atoms. Democritus held that originally the universe was composed of nothing but tiny atoms churning in chaos, until they collided together to form larger units—including the Earth. He believed that a large number of other worlds existed, wandering in space, and that they form and also die continuously, some with life, others dry and barren. He believed that the night sky is filled with stars like our Sun. We will come to the verification of some of these early ideas, but let us start with the discovery of our Earth as a spherical spinning planet.

1. Eratosthenes c276-c194BC

"for demonstrating the Earth is a sphere and measuring its size"

Eratosthenes was born in Cyrene, a city founded by the Greeks in the 7th century BC, but which is now part of Libya. He studied in Athens under the learned philosophers Zeno of Citium and Aristo of Chios. He became known for developing chronology, the scientific study of historical dates, and for his great works of poetry. In 245BC he was invited to work at the great Library of Alexandria and within five years became its chief librarian. He is credited by the great astronomer Hipparchus for inventing the armillary sphere. This is a spherical framework of rings which represent celestial coordinates and is used to map the constellations and the apparent motions of the stars. It was independently invented in China around the same time. Eratosthenes deserves recognition for his visionary experiment to measure the size of the Earth. But before this could be accomplished, there was the difficult step in realising that the Earth was not flat, but a sphere that was spinning in space.

It is not known who first had the idea that the Earth was not flat, or what the evidence was that they used. Certainly, that first step, to move from a flat Earth to a sphere, was bold and brilliant. The evidence for a spherical Earth was available, perhaps the most obvious being the curved shape of Earth's shadow during a lunar eclipse. And there were stories from travellers about a bright star, Canopus, which was not visible in Greece but came into view when one travelled to Rhodes and rose higher in the night sky the further south one travelled. Travellers to the far north had noticed that the length of the day varied, and there were rumours of a place where people slept for six months of the year. An expedition of the great seafaring traders, the Phoenicians, circumnavigated Africa around the 6th century BC. On their three-year journey they passed into the Southern hemisphere and noticed that the Sun was on their right whilst they sailed West.

Hicetas of Syracuse was one of the last philosophers from the Pythagorean school in the 4th century BC. He was perhaps the first to suggest that it was the Earth itself that was spinning on its axis, turning once in every 24 hours. We know very little of Hicetas, just a few mentions including this most important reference from the Roman philosopher Cicero in the 1st century BC: *"Hicetas of Syracuse, as Theophrastus says, holds that the heaven, the Sun, the moon, the stars and in fact all things in the sky remain still, and nothing else in the universe moves, except the earth; but as the earth turns and twists about its axis with extreme swiftness, all the same results follow as if the earth were still and the heaven moved."*

Throughout history creative individuals have been able to find solutions to problems that at first appeared to be very difficult to solve. How would you prove that the Earth is round? How would you measure its size? There is a much simpler way than photographing it from space or travelling all the way around it and returning to where you started. Around 240BC, Eratosthenes accurately measured the size of the Earth and published his results and technique in a book entitled 'On the measure of the Earth'. Although his book has not survived, his methods were described by other Greek and Roman writers.

The later Greek astronomer Cleomedes summarised the techniques that Eratosthenes used along with his estimated size of the Earth, although not the details on how he performed his measurements. Eratosthenes could have made his estimate using a small stick, a calendar and some simple geometry, all without moving from his home. He may have known that on the day of the summer solstice, the reflection of the Sun was visible at the bottom of a deep well in the city of Syene. Or he may have known that a vertical stick in Syene would cast no shadow because the Sun was exactly overhead that day. But at the same time at his home town of Alexandria the stick cast a shadow with an angle of just over seven degrees. This angle represents the difference of latitude of the two sites on the surface of a spherical Earth. Knowing the distance between Alexandria and Syene was 5000 stadia, using the geometry of triangles he would have found that the circumference of the Earth was 250,000 stadia. The value of Eratosthenes stadia was 157.5 metres, therefore his value for the circumference of the Earth was 39,690 kilometres, which is accurate to less than one percent!

Even with our first award, we encounter a discovery that may have been made earlier. Aristotle wrote his famous text 'On the Heavens' around the year 350BC. He describes how the phases of the Moon imply that the Moon is a sphere. He goes on to write about how the positions of the stars change as one moves North or South. Then he mentions the mathematicians who had measured the size of the Earth and quoted a number that was about twice as large as found by Eratosthenes. I wonder who those mathematicians were who first tried to calculate the size of the Earth? Perhaps it was Eudoxus of Cnidus who observed the stars from both Egypt and Greece. Eratosthenes was often called 'beta' by his contemporaries, and some have speculated that this was because he always came second. But this is how science works – step by step on the ideas and findings of others.

Eratosthenes' method for measuring the size of the Earth was simple and accurate. It shows your latitude on Earth and was the basic technique used for geodesy for the next two millenia until satellites were placed into orbit. Eratosthenes became famous for his lost work 'Geographika' in which he described and mapped the entire known world. He divided the Earth into five climate zones and used parallels and meridians to link together all the known cities – essentially inventing geography. He was also a mathematician and invented a method for determining prime numbers. As he aged, he contracted ophthalmia, becoming blind around 194BC. It is said that losing the ability to read and to observe nature depressed him so much that it led to him voluntarily starving himself to death.

2. Aristarchus c310-c230BC

"for discovering our place in the solar system"

For most of human history, it was thought that our Earth was at the centre of the universe. Would you have thought differently without the knowledge you were given? After all, the stars and planets appear to revolve around the Earth and if the Earth were moving wouldn't you surely feel the effects?! And could you think of a way of measuring the distances and sizes of the Sun and Moon without leaving your home? This was accomplished by Aristarchus of Samos. He was also the first to place the Sun in its correct position, at the centre of our solar system, orbited by the Earth and planets.

To understand just how remarkable Aristarchus' measurements and propositions were, we must put it in context with the main ideas on the cosmos at the time. In the 6th century BC Anaximander and his student Anaximenes thought that the Sun, Moon and planets were all made of fire, *"riding on air like leaves because of their breadth"*. They thought the fixed stars were part of a crystal sphere that rotated around the Earth – the beginning of an idea that plagued Western cosmological thought for the next two millennium. Plato knew that the apparent motion of the planets was complicated – at times they seem to move backwards, or retrograde, as seen from Earth. He wished to understand their motions and believed that the cosmos should follow simple geometric forms such as circles and spheres. In the 4th century BC, his student Eudoxus of Cnidus developed the mathematics behind the first geometric model of the cosmos – the idea that 26 perfectly spherical spheres guided the motions of the Sun, Moon and planets. Although Plato and Eudoxus are thought to have been attempting to understand the planetary motions mathematically, these ideas were later refined and transformed into a cosmology by Aristotle.

Aristotle, another student of Plato, developed a celestial model with 55 interconnected moving spheres, centred on Earth and each being moved by its own god. He believed the universe was finite and eternal. The spheres were thought to be made of a transparent 5th element, called the ether or quintessence. An outermost fixed sphere held the stars since they were not observed to move.

The perfectly nested spheres cosmology did not survive for long. It was pointed out that the brightness of Venus and Mars varied over time. We now know this is due to their changing distances from Earth as they orbit the Sun – but if they revolved around the Earth at a fixed distance their brightness should not change. This was explained with a novel modification to the celestial spheres model – that the centres of the spheres holding the planets were offset from the Earth. Hipparchus was the first to develop these ideas mathematically and to test them with observations. The ideas of Aristotle and the mathematical description of the planetary movements were famously taken up by Ptolemy, who added even more crystalline spheres and laid the foundations of astrology as we know it today, helping to set back science a thousand years in the process.

The reason for the additional complexity of the moving spheres was two-fold. Relative to the stars, the planets sometimes appear to move backwards – a retrograde motion that we now know is just due to our changing view of the planets as Earth orbits the Sun. The apparent motion of Mars was particularly complex due to the fact it moves on a rather eccentric orbit, but that was unknown at the time. Even much later, despite correctly placing the planets orbiting the Sun, Copernicus could not match the motion of Mars as he also assumed circular motions.

A motionless Earth around which the Sun and planets revolved was the most obvious viewpoint, and one that existed in every early cosmology. Standing outside and observing the horizon and the night sky, it would seem logical that the Earth is stationary and that everything is moving around us. It was logical to think that the Earth was at the centre of everything since everything appeared to be attracted towards the centre of the Earth.

To an observer in space, we are actually moving at 1,675 kilometres per hour because of the Earth's rotation and 108,000 kilometres per hour due to Earth's orbit around the Sun. That sounds like it should make us dizzy, but there are several reasons why we can't perceive this motion. Earth's orbit is so large that our motion is so gently curved that it feels as though it is motion in a straight line. Motion in a straight line at a constant speed is a natural state of motion, and it is undetectable unless you see something whizzing past you. And because of Earth's rotation, the 'g-force', or centrifugal acceleration, is just 0.003g (or in other words, 0.3 percent the strength of Earth's gravity) and the acceleration from Earth's motion around the Sun is even smaller.

It was Aristarchus who was the first to completely revise our place in the cosmos. Little is known of his life except that he spent at least some years at the museums and libraries in Alexandria. Aristarchus was a student of Strato of Lampsacus, who was the head of Aristotle's Lyceum. According to Archimedes, Aristarchus made several of the most notable contributions to astronomy in ancient Greece towards the end of the 3rd century BC.[1] The only surviving text is his work 'On the Sizes and Distances of the Sun and Moon'.

In this text Aristarchus begins by calculating the relative distances and sizes of the Sun and the Moon. He makes the correct premise that the Moon reflects the light from the Sun and that when the Moon is exactly half illuminated, it must form a right-angled triangle with the Earth and the Sun with the Moon at the vertex of the right-angle. Aristarchus only needed to measure the angle between the Moon and Sun as viewed from Earth. He then used the geometry of triangles developed by Euclid to determine the relative lengths of two sides of the triangle, the ratio of the Moon to Sun distance.

With a second method Aristarchus determined both the absolute sizes and distances to the Sun and Moon. He used the geometry of a lunar eclipse to calculate the radii of the Moon and Sun in terms of the Earth's radius. This second calculation involves a many more triangles and geometry, but the basic idea is that he could estimate the size of the shadow of the Earth at the distance of the Moon, and the size of the Moon relative to Earth's shadow. He could do this by timing how long it took the Moon to cross the shadow (the length of totality during a lunar eclipse) and knowing the orbital time of the Moon about the Earth. Aristarchus also reasoned that since the angular size of the Sun and Moon were the same (by using the fact that during a solar eclipse the Moon almost perfectly fits on top of the Sun), then simple geometry gives the size of the Sun in relation to the size of the Moon.

Aristarchus' measurements were not very accurate, but his methods were correct. He inferred that the diameter of the

---

[1] 'Aristarchus's On the Sizes and Distances of the Sun and the Moon: Greek and Arabic Texts', 2007, J.L. Berggren and N. Sidoli, Archive for History of the Exact Sciences, 61, 3, 213.

Earth was nearly three times the diameter of the Moon and that the Sun was far larger than the Earth. He estimated that the distance to the Sun was about 19 times the distance to the Moon, a value twenty times too small. The Greek astronomer Hipparchus used these techniques to measure a lunar distance and size that were accurate to better than ten percent. Over two thousand years ago it was known that the Earth was three and half times the size of the Moon, which lies at a distance of about 60 Earth radii away[2].

This was an incredible achievement and was the foundation of the 'distance ladder' that enables us to bridge the gaps in measurement techniques to ever more distant objects in the universe. Aristarchus also realised that the stars in the night sky must be much more distant than our Sun since they showed no detectable parallax. Parallax is a technique which uses the change in apparent position of an object viewed from two different places to measure its distance. To see how it works, hold up your finger and look at how the position of your finger changes relative to distant objects as you look with each eye. Now move your finger twice as close and the distant object appears to move twice as far from side to side. This change in apparent position can be measured as an angle. Then all you need to know is the separation of your eyes, and you can measure the distance to your finger using geometry.

---

[2] The difficulty in making accurate measurements of the angles led to uncertain estimates of the size of the Sun and its distance from Earth – Hipparchus found that the Sun was 2490 Earth radii away, but the correct distance is almost ten times larger. We will hear later about how this distance and size of the Sun was accurately determined.

Another visionary idea of Aristarchus is preserved in the text 'Psammites', also known as 'The Sand Reckoner', written by his younger contemporary Archimedes. After describing the prevailing geocentric model he goes on to write that Aristarchus of Samos had published the hypothesis that it is the Sun and fixed stars that are immovable and that the Earth and planets move around the Sun. Aristarchus correctly identified our place in the solar system and proposed the heliocentric model over one and half thousand years before Copernicus in the 16th century AD. He was ahead of his time but did not gain recognition until long after Copernicus.

Archimedes was certainly inspired by Aristarchus – his short text was one of the first 'popular science' works in which he used the results of Aristarchus to calculate how many grains of sand it would take to fill the known universe! But aside from a few citations by other Greek philosophers, it seems the idea did not gather much support. One exception is the Chaldean astronomer Seleucus of Seleucia who supported and taught the heliocentric model of Aristarchus as well as the 24-hour rotation of the Earth. Seleucus is known from the writings of Plutarch, Aetius, Strabo, and Muhammad ibn Zakariya al-Razi. The Greek geographer Strabo lists Seleucus as one of the four most influential astronomers who came from Hellenistic period, around 150BC. According to Plutarch, Seleucus even proved the heliocentric system through reasoning, though it is not known which arguments he used.

The major contributions of the ancient Greeks to the discovery of our universe ended with Hipparchus. He is often quoted as being the greatest astronomer prior to the invention of the telescope. He constructed the first accurate and comprehensive catalogue of the visible stars and discovered the precession, the wobbling axis, of the Earth. He also made mathematical models for the motions of the Sun, Moon and planets. It is highly likely that he designed or even constructed the famous Antikythera mechanism – a remarkable mechanical device a thousand years ahead of its time, with cogs and gears that could predict the future positions of the Moon and planets.

Our journey of discovery continues with the work of the Polish mathematician and astronomer, Nicolaus Copernicus[3].

---

[3] Wait a minute professor, you just skipped one and half thousand years of Western scientific history! That is true, but there is not one single development relevant to the discovery of our universe that comes from this period. Why is that you may rightfully ask? Depending on who you ask you will get a different tale. I put this scientific dark age down to the overwhelming influence of superstition, astrology and dogmatic christianity but others may disagree.

3. Nicolaus Copernicus (1473-1543)

"for the heliocentric model of our solar system"

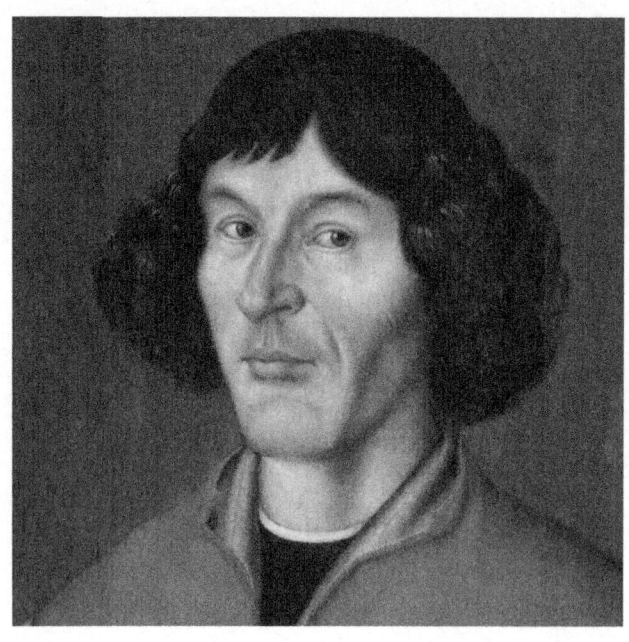

The ideas formulated by Aristarchus were finally recognised and placed on a firm mathematical footing by the Polish mathematician and astronomer Nicolaus Copernicus in 1543 in his major work 'On the Revolutions of the Celestial Spheres'. This began the path of placing the Sun back in its rightful place at the centre of our solar system and prompted people to start discussing astronomy and science again.

The 15th century marks the beginning of the renaissance in Europe epitomised by the free-thinking creative genius of Leonardo da Vinci. Johannes Gutenberg invented the printing press that enabled the wider dissemination of news and knowledge. The Byzantine Roman Empire had fallen to the Ottoman Turks and many scholars fled to the West, bringing with them more of the ancient Greek texts. The second scientific revolution was about to begin and its starting date is often set at the publication of the work of Copernicus.

Copernicus was born in 1473 in Royal Prussia, which is a region that had been part of the Kingdom of Poland since 1466. Whilst studying at the University of Kraków, Copernicus learned astronomy and mathematics from the writings of the ancient Greeks, such as Aristotle and Euclid. At this time in history, astrology and astronomy were considered subdivisions of a common subject called the 'science of the stars', whose main aim was to provide a description of the arrangement of the heavens, particularly for navigation and calendar keeping, as well as the theoretical tools and tables of motions that would permit accurate construction of horoscopes for making prophecies.

Later, Copernicus moved to Bologna to study for a degree in canon law. Whilst there he shared a house with the principal astronomer at the university, Domenico Maria de

Novara. He became his assistant, helping him in his chief role of issuing annual astrological prognostications for the city, and in particular the fate of the Italian nobility and their enemies. Copernicus became familiar with the works of Ptolemy as well as the main critiques. These included 'Disputations against Divinatory Astrology' by Giovanni Pico della Mirandola which pointed out that astronomers not only disagreed on the division of the zodiac, but they also disagreed about the order of the distances to the planets, therefore astrologers could not be certain about the strength of the power of these planets on Earth.

In 1503 Copernicus received his doctorate in canon law and worked for the church, but continued astronomy in his spare time for which he gained a widespread reputation. In 1514 he wrote a brief outline of his ideas where he postulated that if the Sun were at rest and orbited by the Earth and planets, then their orbital periods would increase with their distance from the Sun. This text is known as the 'Little Commentary' and contains seven postulates, in which he states that the planets rotate around the Sun and the Moon rotates around the Earth. The seventh postulate was the statement that the apparent retrograde motion of the planets is an illusion that results from viewing their positions from a moving Earth.

It took another 20 years for Copernicus to work out the details of his model and to complete his famous manuscript 'Dē revolutionibus orbium coelestium'. It explains in mathematical detail how the apparent motions of the planets could be understood if they orbited a stationary Sun. Copernicus delayed publishing his work to avoid the wrath of the church. However, rumours of its contents spread widely over Europe, and he was eventually persuaded to publish. Copernicus died at the age of 70 in 1543 and according to legend, he was presented with the final printed version of his work on the very day that he died.

A fundamental principle of cosmology today is the so-called 'Copernican principle', which states that the Earth is not in a central, special place in the universe. Perhaps this should be called the Aristarchus-Copernican principle. Copernicus was aware of the ancient Greeks who thought the Earth moved, referencing Aristarchus in an early draft of his book, but the reference was removed from his final version[4].

The view of our solar system advocated by Copernicus did not gain immediate acceptance, even by the astronomers at the time. There were arguments about how the air on a spinning Earth would blow the buildings down, how stones falling from a tall tower should land a long way from its base. Another complaint was why should all the celestial bodies orbit the Sun apart from the Moon that orbited the Earth? At the time, the satellite moons of Jupiter and Saturn had not yet been discovered.

Aristotle's writings about a stationary Earth had left their legacy, along with a single sentence in the bible that was interpreted to mean that the Sun revolved around the Earth. Upon hearing about Copernicus' work, Martin Luther commented *"People give ear to an upstart astrologer who strove to*

---

[4] We will encounter 'Stigler's law of eponymy' many times in this book. Statistics professor Stephen Stigler stated in 1980 that no scientific discovery is named after its original discoverer. He quoted examples such as the Pythagorean theorem, which was known to Babylonian mathematicians. Even his own law, he quotes, had been noted by others! Although Copernicus may only have been aware of the reference to Aristarchus by Plutarch in which it is only mentioned that *"Aristarchus thought the Earth moves"*.

show that the earth revolves, not the heavens or the firmament, the sun and the moon ... This fool wishes to reverse the entire science of astronomy; but sacred Scripture tells us [Joshua 10:13] that Joshua commanded the sun to stand still, and not the earth."

4. Johannes Kepler (1571-1630)

"for the laws of planetary motion"

Little support was given to the work of Copernicus after his death. That changed thanks to Johannes Kepler who was born 28 years after Copernicus died in the Free Imperial City of Weil der Stadt, now in Germany. Kepler was the first scientist to discover universal laws of nature.

Kepler had an unhappy childhood and wrote that his father was an immoral and quarrelsome soldier and that his mother was generally an unpleasant person. However, he recalls an early interest in astronomy, when at the age of six, he saw the great comet of 1577. This was the same comet that the Danish astronomer Tycho Brahe was observing and found that it must move beyond the 'crystaline sphere' of the Moon. The lives and works of Brahe and Kepler later became closely entwined and it was the careful astronomical observations of Brahe that led to Kepler's greatest discovery.

Since he excelled at school Kepler was able to enter the Protestant Tubingen University with the intention of becoming a clergyman. Whilst there he was taught by Michael Maestlin who was professor of mathematics and astronomy. Maestlin was required to teach the geocentric astronomy of Aristotle and Ptolemy. However, in private, he also taught the ideas of Copernicus which were simpler and more intuitive. Kepler was intrigued and this began him on his path of scientific discovery. He went on to become a mathematician of the state in Graz and began his lifelong quest to understand the motion of the planets.

Kepler began on a rather mystical path, trying to fit the presumed circular orbits of the planets within the five 'perfect three-dimensional shapes', originally associated by Plato to the Earth, air, water and fire. These shapes, which include the four-sided tetrahedron and six-sided cube, are the only five shapes that have faces identical in shape and size, which connect with the same angles and with the same number of faces meeting at each vertex.

Kepler published his theories in the 1596 book 'Mysterium Cosmographicum' which was the first work to defend the ideas of Copernicus. Despite the mystical approach, there was also brilliance in the text. He argued that Brahe's observation of the comet beyond the Moon excluded the possibility of crystalline shells carrying the planets. He was the first to suggest that a physical phenomenon, a radiated force associated with the Sun, must be responsible for the planetary movements. Kepler also wrote that a force emanating from the Sun that decreased with distance might account for the decrease in orbital speeds of planets further from the Sun.

Kepler's ideas on perfect harmony between shapes and the planetary orbits did not quite work out. He thought it was because his data was not very accurate, but he knew that Tycho Brahe had more accurate observations of the planets that he needed to test his theories. In the year 1600 the conflict between protestants and catholics was reaching a peak and it looked unlikely that Kepler could stay in Graz. Kepler travelled to Prague to secure employment from Brahe. It was the first known collaboration between an observational astronomer and a theoretician. Brahe wanted Kepler to verify his own Earth-centric model, and Kepler wanted Brahe's data to verify his harmonious planetary orbits about the Sun.

Tycho Brahe also rejected the ideas of Copernicus of a moving Earth and came up with his own model of the planetary motions. In his model the Earth was placed back at the centre of the universe, orbited by the Moon and Sun, whilst the five planets orbited the Sun. Unknown to Brahe, a similar model had been developed earlier by the Indian astronomer Nilakantha Somayaji[5].

Despite his rejection of Copernicus, it was Tycho Brahe's observations that ultimately led to the demise of the crystalline shells and Earth-centrism. Brahe was a wealthy nobleman, as were many of the first scientists. They had the time and money to study, contemplate and experiment. Brahe built an astronomical observatory from which he systematically studied the night sky, keeping careful notes on the positions of the planets. In 1572 he observed the appearance of a new star – a supernova. Brahe tried to measure its distance using parallax, but it showed none, implying that it must be further away than the Moon and the planets. He also observed the great comet of 1577 crossing the celestial sphere – the impenetrable crystalline shells. That this holy realm of god was not invariant was an idea very much disliked by the church which insisted the outer celestial sphere that hosted the stars and the heavens was immovable and unchangeable.

Brahe was also as eccentric as many scientists today. He lost part of his nose in a duel with another student after arguing over who was the better mathematician. He hosted extravagant parties during which his guests would be entertained by his pet elk, which died one night after drinking too much beer and falling down the stairs! However, Brahe was rather protective of his astronomical data and did not even allow his new assistant, Johannes Kepler, full access to his notes.

---

[5] It was described in his Aryabhatiya-bhasya, a commentary on the texts of the famous Indian astronomer Aryabhata (476-550). He also thought that the apparent motions of the stars could be understood if the Earth itself rotated.

Brahe died in 1601 and on his deathbed, he asked Kepler to continue the reform of astronomy, but on the basis of his own models rather than those of Copernicus. However, Kepler thought the model of Copernicus made much more sense. He carefully studied Brahe's data on the movement of Mars and supplemented them with new observations of his own. Kepler found that the movement of Mars did not match his presumed circular motion and he had to reject his, and all previous notions of its orbit.

Kepler thought that this might be due to the assumption of Copernicus that the Earth moved at a constant speed around the Sun. "What if the Earth's motion were not uniform", he asked? To test this Kepler came up with an idea, described later by Einstein as an idea of a true genius. Using a triangulation procedure, Kepler calculated Earth's orbit as it would be seen by an observer on Mars. He discovered that the Earth did not move at a constant speed and that the Sun is not at the centre of Earth's orbit. This led him to the observation that the planets sweep out equal areas in equal times, one of his three famous laws of planetary motion. Then he considered the motion of Mars and came up with his most famous law of planetary motion – that its orbit is an oval, an ellipse, with the Sun at one focus of the ellipse.

In 1608 Kepler published his most famous work, 'Astronomia Nova', in which he showed that Mars lies on a precisely repeating but non-circular elliptical orbit about the Sun. The crystalline spheres and circular motions, that had been assumed by all astronomers since the time of Plato, were becoming cracked and faulted. He went on to argue that it was a force from the Earth that kept the Moon in motion and that a similar force from the Moon was responsible for the ocean tides on Earth. Later, he discovered his third law of planetary motion, which relates the orbital period of a planet to its distance from the Sun. Isaac Newton would later become famous for explaining the origin of these laws under a universal law of gravitation.

Kepler's laws of planetary motion were the first scientific 'laws of nature' to be discovered. These are rules that can be used to predict the behaviour of a physical system in time and space. Although his theories were not immediately accepted, he had used his new laws of planetary motion to predict that Mercury would transit across the face of the Sun in 1631. This was duly observed on the exact day he had predicted, but unfortunately Kepler died the year before and did not see his work verified.

5. Galileo Galilei (1564-1642)

"for the evidence that the planets orbited the Sun and for the discovery of our galaxy"

The final nail in the coffin of the celestial spheres came thanks to the Italian astronomer Galileo Galilei, one of the first scientists in the modern sense of the word. Encouraged by his musician parents, Galileo began studying medicine at the university of Pisa, although he attended lectures in mathematics and natural philosophy which were his real passion. Whilst a student he noticed a swinging chandelier and used his pulse to measure that it took the same amount of time to swing back and forth no matter how far it was swinging. He began experimenting with pendulums to verify his observation and showed that the time of swing of a pendulum depends only on the length of the string. A hundred years later, Christiaan Huygens would use this principle to create the first pendulum clock. Galileo left university without a degree and taught privately whilst carrying out his experiments. In 1859 he became a mathematics professor at the University of Pisa on a temporary teaching contract.

Galileo tried to explain phenomena with mathematics, changing natural philosophy from a subject that was qualitative to one that was quantitative and could be tested by experimentation. One example you may have heard was his experiments dropping objects of different weights from the leaning tower of Pisa. He showed that Aristotle's ideas on motion were incorrect, that all objects fall at the same speed independent of their mass. He used logic to argue that motion is relative – that just because the Earth is moving, we could not measure that motion or feel its effects. This is now called 'Galileo's principle of relativity' and predates one of the key ideas of the theory of special relativity by several centuries. Galileo was perhaps the first scientist who resembled the scientists of today – breaking free from the old ways of thinking. His way of approaching science predated the famous exposition on the scientific method by the English philosopher Francis Bacon in 1620.

Galileo's attacks on Aristotle were not so popular amongst his colleagues and his contract at Pisa was not renewed. He moved to the University of Padua where he became chair of mathematics from 1592 until 1610. After his father died, Galileo, the eldest of six children, was responsible for supporting his family but his university salary was not large and their family was not wealthy. He manufactured his inventions in his spare time to sell for extra money. One of these was the thermoscope – a device for measuring temperature by monitoring changes in the density of air. This would later evolve into the modern thermometer.

In 1609 he showed that the distance fallen by a body is proportional to the square of the time taken to fall (this means an object will fall four times further in two seconds than in one second), and that projectiles follow parabolic orbits. Both of these results again contradicting Aristotelian physics. But it was not until 1610, at the age of 46, that he made his most famous discoveries. These were enabled thanks to one of the greatest inventions in human history - the telescope.

The ancient Greeks knew how to make glass lenses that could be used to focus the light of our star to light fires. Spectacles were used in Europe as early as the 13th century. But it puzzles me that it was not until the beginning of the 17th century that it was realised that two lenses held in front of each other made distant objects much easier to see. This is the basic principle for the workings of a telescope, and it is thought to have been invented by the German-Dutch spectacle-maker Hans Lipperhey in 1608.

The furthest convex lens collects light from a larger area than that possible with your eye and focuses that light onto a second lens which refocuses the light. What you see is a magnified image with more clarity. Two convex lenses used like this will make a distant object appear upside down, but if you use a concave lens for the second lens then the image will be the right way up. How simple is that?

It is this simple invention that began the path of understanding our cosmos. It is one of the greatest inventions of our species, allowing us to observe the distant and the small - a microscope works in the same way as the telescope, the only difference is how far apart you put your two lenses. Without the telescope and microscope, I suspect that we would still be in the dark ages today.

The credit for the first person to take two lenses and hold them apart is given to Lipperhey. Lipperhey was born in the region that is now Western Germany. He later became a citizen of Zeeland, now part of the Netherlands, where he ground lenses and made spectacles. In 1608 he applied for a patent for his invention for seeing things far away as if they were nearby. However, a few weeks later, another Dutch lens maker, Jacobus Metius, filed a similar patent. Metius had already heard of Lipperhey's application and quicky filed his own claiming that he had discovered it over a year before. The state awarded neither a patent, since it was deemed too easy to copy, but they credited Lipperhey and paid him well to produce several telescopic devices.

News of this invention quickly reached Galileo, who worked out its design from the limited knowledge he had. By the end of the summer in 1609 Galileo had learned how to grind his own lenses and had made telescopes with a magnifying power of nine which was several times more powerful as achieved by Lipperhey. Upon showing his improved telescope to the Venetian Senate, Galileo was awarded with lifelong tenure and his salary was doubled. By the end of the year Galileo had produced telescopes that had a magnifying power of over twenty. He pointed his telescope at the night sky and began observing the planets and stars.

Galileo found that the fuzzy band of light that is our galaxy, 'The Milky Way', was actually made up of countless stars confirming the ideas of Democritus. He was the first to observe the craters on the Moon showing it was not a smooth sphere, but was covered with craters and mountains. It was as if a new world had been discovered and the public flocked to see its surface. This was another blow to the supporters of Aristotelian philosophy that required the heavenly bodies to be pure and perfect spheres. He discovered the brightest four moons of Jupiter, we now call them 'the Galilean moons', and he would later observe their orbital motion. Previously it was argued that if the Earth moved, it would leave the Moon behind, but this was a clear counter example. Galileo was the first to observe the motion of sunspots showing that the Sun rotates. He published many of his findings in 'Sidereus Nuncius' in 1610 which made him a scientific superstar overnight.

He turned his telescope towards Venus and saw that it was half illuminated – he found that Venus passed through phases rather like the Moon! Over the next months as Venus moved closer to the Sun, the illuminated part grew larger until Venus became a full disk of light. Galileo had discovered that Venus was also a spherical object illuminated by the light of the Sun. When the Moon is close to the Sun, it lies between the Earth and the Sun and we only see a partly illuminated crescent. But because Venus orbits around the Sun, from our perspective it reaches full illumination when it appears closest in the sky to the Sun. The only explanation of the phases of Venus relative to the Sun was that Venus was orbiting the Sun and not the Earth. Galileo had found the first evidence in support of the heliocentric model of Copernicus. If Venus had been a larger or closer world then the ancient Greeks could have proven the heliocentric model 2000 years ago because they would have witnessed its change of phases by eye.

Unfortunately, news of Galileo's findings upset the church which was insistent on the literal interpretation of the bible – the Earth should not move! In 1615 the Inquisition of the catholic church pronounced the Copernican theory heretical and banned its teaching. In 1616 church officials were asked to decide whether to accept the notion that the Earth moves. Accordingly, Galileo was summoned to Rome to examine the pros and cons of heliocentric astronomy. The church was willing to accept the heliocentric model as a calculating device, but not as literal truth. Pope Urban VIII initially encouraged Galileo to publish his arguments for and against heliocentric astronomy, but quickly regretted this decision since Galileo could not refute his own findings. He was arrested for ostensibly making fun of the Pope in 1633. He was put on trial for his ideas, found guilty and was placed under house arrest at his villa in Arcetri near Florence until his death in 1642. It was not until 1992 that the catholic church revoked its condemnation of Galileo and apologised for his mistreatment.

The works of Copernicus, Kepler and Galileo were placed on the catholic churches' Index of Forbidden Books for teaching that the Earth moves, contrary to holy scripture. But science is not about belief. Science is based on observation, experimentation and the testing of theories with reproducible evidence, and the evidence could not be ignored.

The scientific revolution began in earnest in Europe in the 17th century which saw the gatekeepers of knowledge move from the church and universities to scientific societies. The public were also fascinated by the new emerging knowledge and lecturers of science travelled across Europe explaining the results to enthralled audiences. The grip of religion, that had held Europe in its control since the Middle Ages, began to decline. However, it was still demanded by Western scholars and universities that nature be interpreted as a consequence of god's actions and scientists had to interpret their findings within a biblical context. Our next brilliant individual was no exception.

6. Isaac Newton (1642-1726)

"for demonstrating that a force called gravity gives rise to elliptical planetary orbits"

The English scientist Isaac Newton is often listed alongside Albert Einstein as one of the most influential scientists of all time. From inventing the first reflecting telescope that used a mirror instead of a glass lens, to discovering the colours of sunlight using a prism, he made numerous discoveries that warrant this claim. His most famous work was 'Principia', in which he established the laws of motion and gravitation. He was born the same day that Galileo died. Whereas Kepler and Galileo had discovered how things move, Newton would go on to discover why they moved.

Newton had a turbulent childhood that may have led to his well-known psychotic tendencies. He was prematurely born into a middle-class English family, but his father died a few months before he was born. He was encouraged to become a farmer but his school teacher persuaded his mother to let him study. At Cambridge University he read Aristotle's works, like all students at the time. But he also delved into the publications of scientists such as Kepler, Galileo and Descartes. Whilst he was a student he invented calculus, but then in 1665 the plague hit England and the universities were closed. It was during this two-year 'lockdown' period at home that he developed his ideas on motion and gravitation.

Newton's laws of motion sound simple, but they had puzzled philosophers for thousands of years. According to early christianity, the planets were carried by angels and spirits based on Aristotle's erroneous notions that whatever moves must be moved and that constant motion requires a constant force. In Aristotelian physics, weight was an inherent property of an object, and its natural state was to be at rest. These early incorrect ideas on matter and motion were based on naive observation, but perhaps one that most of us would also make based on our everyday experience.

The correct understanding of how objects move began with Galileo, who performed experiments on the motion of bodies to infer that they always fell to Earth, accelerating at a rate equal to 9.8 metres per second each second independent of their mass. Galileo also discovered the fundamental concept of 'inertia', the tendency of matter to preserve its state, whether it be at rest or in motion. The French scientist René Descartes generalised this concept to assert that the natural state of motion of an undisturbed body in the universe is not at rest, or in circular motion, but moving in a straight line at constant speed. This was the foundation of mechanics! Without these insights, Isaac Newton's famous laws of motion could not follow.

The principle of inertia, or Newton's first law, states that if something is moving with nothing touching or disturbing it, it will keep on moving at the same speed in a straight line forever. Newton's second law describes how one must exert a force to change the speed or direction of motion of an object. Newton's third law is the action-reaction principle, of cause and effect. It was not until Emmy Noether's work in the early 20th century that it was revealed why nature can be described by such laws, we will meet Noether again later in this book.

Newton's first law seems strange to us since whenever we set something in motion it never travels in a straight line and it eventually comes to a standstill. The reason is that the force of gravity pulls the object towards the centre of the Earth, causing its trajectory to curve, and friction against the air molecules that are constantly bouncing off its surface will cause it to slow down and stop. The total energy is still conserved, the energy in the motion of the object has been transferred to the motions of the molecules of air, which become hotter. The nature of inertia and Newton's laws of motion are neither intuitive nor obvious. If they were, persistent thinkers such as Plato and Aristotle might have discovered the correct laws of nature long before.

I think that Newton's greatest achievement was to show mathematically that a gravitational force would naturally give rise to planets moving on elliptical orbits as discovered by Kepler.

According to William Stukeley, who wrote a biography of Newton in 1752, Newton really did come up with his idea of a gravitational force after watching an apple fall to Earth: *"After dinner, the weather being warm, we went into the garden, and drank thea under the shade of some apple trees;...he told me, he was just in the same situation, as when formerly, the notion of gravitation came into his mind. Why should that apple always descend perpendicularly to the ground, thought he to himself; occasion'd by the fall of an apple, as he sat in contemplative mood."* However, the idea of such a force had already been contemplated by others, including Kepler.

Newton worked on the assumption that the same mysterious invisible force that caused the apple to fall, must act between the Earth and Moon to keep it in its motion about the Earth. Otherwise, it would travel in a straight line away from us. The force of gravity from Earth must cause the Moon to fall back to Earth, but at just the right amount to compensate for the distance it moves away from the Earth in the same instant of time. And it does. How can it be constantly falling without getting any closer? This is easier to see if you imagine dropping a ball and measuring the time it takes to hit the ground. Now throw the ball horizontally above the ground. It will still fall to the ground in the same time no matter how fast you can throw it, since the gravitational force from the Earth only affects the ball perpendicular to the surface of the Earth. However, if you could propel the ball fast enough, at about eight kilometres per second, the Earth's surface will curve away from the ball at the same rate at which the ball falls back to Earth. In the absence of friction from the air, the ball would continue to travel around and around the Earth, constantly falling but moving forwards quickly enough that it would stay the same height above the Earth's surface.

Newton proved that the strength of this force decreases as one over the square of the distance. If two objects are moved twice as far apart, the force between them becomes four times weaker. When they are three times further apart, the force becomes nine times weaker. Using the inverse square law of gravity, we can calculate how far the Moon must fall in one second knowing that the acceleration must be weaker by a factor which is the square of the ratio of the Earth's radius to the distance between the Earth and the Moon. We find that the Moon falls about one millimetre towards the Earth every second, during which time the Moon has moved forward one kilometre. Because the Moon orbits the Earth about 385,000 kilometres away, this is exactly the distance it needs to fall towards Earth each second so that it stays the same distance away from us. When Newton first did this calculation, the numbers did not quite work out precisely enough for him. Disappointed, he delayed publication of his Principia for several years, until a more accurate measurement of the size of the Earth was made and his predictions turned out to be correct.

Newton asked the question, what if all bodies attract each other with an invisible force? What a question. And what an answer he found! His work gave us the first predictive framework within which we could begin to answer those long-standing questions about our origins. His masterpiece, Principia, was three volumes published in Latin in 1687. They contained many other new results, from estimates of the relative masses of the planets to the distortion of Earth's shape due its rotation to the mechanism that generates the ocean tides. The success of Newton's theories heralded the beginning of modern predictive science. Whilst still on friendly terms with the English scientist Robert Hooke, he had written in a letter *"If I have seen a little further it is by standing on the shoulders of giants."* A rare moment for Newton who later tried hard to play down the contributions of others, often removing all references to the work of scientists with whom he had conflicts.

Isaac Newton could not accept criticism and held furious debates with leading scientists of the time. Most famously having repeated arguments with the German mathematician Gottfried Leibniz about who invented calculus, and with Hooke about the correct interpretation of light and who came up with the law of gravitation first. In fact, in a 1679 letter to Newton, Hooke had suggested to Newton that he should try an inverse square force. Even earlier, in 1645, the French astronomer Ismaël Bullialdus had proposed this law. However, it was Newton who proved it was the only force law that could give rise to the planetary orbits.

In the Principia Newton did not discuss how the mysterious force of gravity worked. But in private correspondence with his colleague, theologian Richard Bentley, he considered gravity to be an instantaneous action of divine intervention via an immaterial ether which Newton identified as due to god. Newton believed that the mechanistic movement of the planets was like a clock that continued to be driven by the clockmaker rather than by its components.

Newton was very much against gravity being due to action at a distance because of some innate property of matter. In 1692 he wrote in a letter to Bentley: *"That one body may act upon another at a distance through a vacuum, without the mediation of anything else, by and through which their action and force may be conveyed from one another, is to me so great an absurdity, that I believe no man, who has in philosophical matters a competent faculty of thinking, can ever fall into it."*[6]

One could fill a book with the fascinating debates that took place in the 17th and 18th centuries about god versus action at a distance and the nature of gravity. In an unpublished manuscript from 1666, Newton wrote that a divine power coexists with space, which extends infinitely in all directions and is eternal in duration. Newton's universe was the sea of stars that we see in the night sky, and he realised that this was an unstable situation under an attractive law of gravity. After all, Newton knew that when apples depart from their branches, they fall to Earth. A static distribution of stars would be unbalanced and collapse together owing to their mutual gravitational attractions. He wrote in the Principia: *"And lest the systems of the fixed stars should, by their gravity, fall on each other, he [god] hath placed those systems at immense distances from one another."*[7] He argued that the starry heavens must be infinite in extent, if it were finite then gravity would cause

---

[6] Letter from Isaac Newton to Richard Bentley, held at Trinity College Library, Cambridge

[7] The mathematical principles of natural philosophy, Isaac Newton 1846 (originally 'Philosophiæ Naturalis Principia Mathematica' pub. 1687 in Latin), translated to English by Andrew Motte, pub. Daniel Adee, New York, page 35.

everything to collapse into one great spherical mass. But he also realised that even in an infinite universe, if the force of gravity between the stars was not perfectly balanced, it would ultimately collapse. He took this as evidence for a god who had arranged this perfect equilibrium.

I wonder why Newton never calculated the time it would take for his unstable universe of stars to collapse. At the time, there were lower limits to the distances to the stars from the lack of a detectable parallax. A simple calculation based on the large lower-limit distances to the stars would have yielded an age of at least several million years. At the time, it was widely accepted that the Earth was about 6000 years old, a number many people had calculated from a literal interpretation of the bible. Newton is widely quoted as also calculating this age, but I can find no record of this in his writings. Newton had calculated that it would have taken the Earth at least 50,000 years to cool from an initially molten hot state, a number he estimated by comparing with how long it took a small iron sphere to cool. But he did not associate this with the age of the Earth. Perhaps he wished to avoid the obvious conflict that would have resulted by discussing such lengthy and disparate timescales.

In his later years Newton often withdrew for years at a time to study alchemy, his secret passion. He was convinced he could make gold from other elements and experimented with lead and mercury, often tasting the products of his experiments. In 1979 a hair found in one of his notebooks was analysed and found to contain lead and mercury at dangerously high levels. Some historians have argued that this may have been responsible for his breakdowns and sometimes obnoxious behaviour.

7. Edmund Halley (1656-1742)

"for shattering the crystalline spheres and for the technique for measuring the distance to the Sun"

By the 17th century, the distance to the Sun was still based on the techniques pioneered by Aristarchus which were not very accurate. But how else can we measure distances to the Sun and stars beyond where we have already travelled? It was the English astronomer Edmund Halley who came up with a simple method of measuring the distance to the Sun with a much higher precision.

Edmund[8] Halley was born into a family that had made its wealth from making soap – a phenomenon that only became widespread in Europe during the 17th century[9]. Despite losing much of their fortune in the great fire of London, they could still afford a good education for their son. Halley became fascinated with astronomy at a young age. Aged 17 he began his studies at Oxford University and assisted the Astronomer Royal John Flamsteed. Halley quickly became an influential scientist who became the second Astronomer Royal and who made many important discoveries in cartography and astronomy.

Halley is most often associated with 'Halley's Comet', the comet he was the first to observe in 1682, and he also realised that it was the same comet that had been sighted in 1607 and 1531! Moreover, he calculated that its orbit is an ellipse, but perturbed by its passing close to Jupiter. It was a grand confirmation of Newton's ideas. Unfortunately, he died before his predicted return of this comet in 1758. Halley is credited with the first accurate eclipse map showing the path of the Moon's shadow across England during the total eclipse of

---

[8] Often called Edmond Halley by biographers, Halley himself used 'Edmond' in three of his publications and 'Edmund' in 22, so I will stick with his preference.

[9] 'Edmond Halley and his Times', S. Jones 1947, Popular Astronomy, Vol. 55, p2-17.

1715. He knew that eclipses seem to repeat in cycles of 18 years plus 10 or 11 days (depending on the number of intervening leap years). Halley named this the 'Saros cycle'. It was first discovered by Babylonian astronomers, and was used for lunar eclipse predictions. Using Newton's theory of the Moon's motion and the Saros cycle, Halley made a series of calculations to identify ancient eclipses in the literature. But Halley soon encountered a problem. The eclipse paths he predicted were shifted with respect to the historical records. Either the Moon was accelerating in its orbit or Earth's rotation rate was slowing down (implying that the length of the day was increasing). Although both are true, Halley correctly identified the increasing length of the day as the primary culprit.

In 1717 Halley was the first to observe that the stars move and are not fixed in crystalline spheres. He measured the positions of the stars Arcturus, Sirius, Betelgeuse and Aldebaran, and compared the values with those measured by Hipparchus and Ptolemy. He found that all the stars had moved! One of the stars had moved half a degree, about the diameter of the full Moon, over 1800 years. This was a remarkable discovery, although later astronomers argued that Halley may have overestimated the accuracy of Ptolemy's data and his results for several of the stars are thought to be spurious. Within a few decades other astronomers had confirmed that some of the stars show movement in the skies and the ancient ideas of a fixed crystalline shell holding the stars was shattered. However, perhaps his most important legacy is inspiring the 1761 expeditions to measure the distance to the Sun.

In 1677 Halley travelled to Saint-Helena, a small island in the South Atlantic Ocean. His aim was to observe the stars in the Southern hemisphere to complement Flamsteed's program of surveying the stars in the Northern hemisphere. On the journey there he passed the equator and through measuring the swing of a pendulum confirmed the idea of Isaac Newton that the Earth must be wider at the equator because of its rotation. The swing of a pendulum depends on the gravitational force which varies slightly depending on how far from the centre of the Earth the pendulum is.

That the time of the swing of a pendulum varies with its position on the Earth is because the Earth is not an exact sphere. It is slightly wider at the equator due to its rotation – the centrifugal force causes the distortion because the Earth is not a solid object but mainly molten 'liquid' rock. Therefore the pendulum would take slightly longer to swing if you stood on the equator where the Earth's gravitational force is weaker. Indeed, on the equator your weight would be half a percent less than it would be at the poles.

Halley's pendulum experiments on the equator played a role in the development of a standard unit for measuring length. Until the 18th century, length was rather arbitrarily defined in terms of body parts, such as the length of a king's foot. The architect and astronomer Christopher Wren proposed defining a standard distance measure as the length of a simple pendulum which swings once from left to right in one second. The alternate suggestion at the time was to define the metre as one ten-millionth of the distance from the equator to the North Pole in a line passing through Paris. I know which method I would have chosen! However, the French academy of sciences selected the latter method in 1791, arguing that the force of gravity varies slightly depending on where you stand on Earth's surface[10].

Halley spent a year and a half at Saint-Helena (in the South Atlantic Ocean) and created the first catalogue of the positions of southern stars. Whilst he was there, he observed a transit of the planet Mercury across the Sun. He made careful notes of the time at which Mercury entered and exited the disk of the Sun and measured the times to an accuracy of one second. Halley realised that if a transit were observed simultaneously from different latitudes on Earth, the transit times would be different since they would be seeing Mercury cross the Sun from slightly different angles. Using simple geometry and knowing the size of the Earth, the distance to Mercury could be calculated using the difference in transit times. Then, knowing the relative distances of the Earth and Mercury from the Sun using Kepler's laws, the distance to the Sun from the Earth could be measured.

Halley calculated that this parallax measurement would be easier to accomplish during a transit of Venus across the Sun rather than Mercury, because of the geometry of the transit. In

---

[10] To define the metre, French scientists set out to measure the distance between Barcelona and Dunkirk. From this they could infer the distance from the North Pole to the equator using geometry. It took over six years. Ten-kilometre baselines were established by measuring the exact distance along the straightest roads in Europe using a 3.9 metre platinum rod. This was used to calibrate the curve of the Earth and the larger distances were measured using triangulation. A standard prototype metre length was established, although it was short by a fifth of a millimetre due to the miscalculation of the shape of the Earth. Thankfully, we now have much more accurate techniques by which we can measure length. Since 1983 one metre has been defined as 'the distance that a photon of light travels in a vacuum in 1/299,792,458 of a second'.

1716 Halley proposed that the next transit events should be observed at different places across the planet and the exact times of the transits accurately recorded. Halley was apparently unaware that the idea had already been proposed, but with few details, in 1668 by the Scottish mathematician James Gregory.

The transits of Venus occur about every 100 years in pairs separated by about 8 years. This is because we need to wait for the orbits of Earth and Venus to align, just as we must wait for the alignment of the Earth and Moon with the Sun to see a total eclipse. The next two transits of Venus will occur in 2117 and 2125. Unfortunately for Halley, at that time the next transit was not until 1761. However, organising such a task at this time in history was a complex operation, particularly because of numerous ongoing conflicts between countries, such as the Seven Years War between France and Britain.

In one of the first international collaborative experiments, motivated by Halley, astronomers from across the world set off on long journeys to observe the transit of Venus from widely separated places across the planet. When the event took place, over a hundred astronomers at over 60 locations across the planet were in place ready to observe the event. It was the 18th century equivalent of today's big scientific endeavours such as the Large Hadron Collider at CERN – the first global collaboration of scientists.

Many had travelled for weeks or months to reach their destination. Some never managed to get to where they needed to be. One pair of British astronomers set off sailing to the East Indies, but their ship was attacked by a French warship before it had even left the English Channel, leaving 11 dead. After repairs and being reminded of the British importance of 'keeping a stiff upper lip', the terrified astronomers set sail again.

The tale of French astronomer Guillaume Le Gentil is particularly sad. In 1760 he set sail to India to observe the transit. On the way, however, he learned that a war had begun between France and Britain and thus it was too dangerous to try to reach his destination. He was a determined fellow, so he then headed to an alternative part of India only to find that it was also occupied by the British. His ship was ordered to turn back. Although the sky was clear, when at sea it was impossible to observe the transit because of the motions of the boat. Disappointed, he decided to await the next transit event eight years later since he had already travelled so far. He sailed to the Philippines to observe the 1769 transit but was turned back by the Spanish authorities there. Unperturbed, he headed back to his original destination in India which was back under French control. He constructed his observatory and waited patiently. Even though every day each of the previous weeks had been clear, on the day of the event the skies clouded over and he was unable to see anything. He only just managed to avoid sinking into insanity and started the long journey home to France, with many adventures along the way. He eventually arrived back in Paris 11 years after he had set off to find that he had been declared legally dead, his wife had remarried, and his relatives had shared his estate amongst themselves. What some people will do in the pursuit of scientific discovery!

Luckily, the other expeditions were a lot more successful. Further expeditions were organised to observe the 1769 transit, including Captain Cook's famous three-year voyage to Tahiti. The combined results revealed the distance to the Sun that was just a few percent different from modern radar measurements of 150 million kilometres.

Unfortunately, Halley died 20 years before these famous transits took place. However, his legacy will long live on. For thousands of years, it was thought that comets appeared randomly and were some sort of message of doom from the prevailing gods of the time. Halley showed that comets were just like the planets in our solar system, objects orbiting our Sun following predictable paths. He showed that Halley's comet will continue to visit us, approximately every 76 years, ensuring that his name will be remembered every time the comet visits the inner solar system, long after most scientists' names are forgotten.

8. Immanuel Kant (1724-1804)

"for his visionary ideas of the origin of the solar system and of a vast and evolving universe"

By the mid-18th century, ideas about cosmology were no better than those of the ancient Babylonians, with spherical shells of stars orbiting a divine centre. This all changed with the insightful brilliance of the German philosopher Immanuel Kant.

Kant was born in Königsberg near the Baltic Sea. Then German speaking and part of Prussia, it is now called Kaliningrad and is part of Russia. While at university Kant began to study philosophy and mathematics, and wrote his first published work at age 23 trying to resolve the philosophical dispute between the followers of Newton and Leibniz about the nature of matter. He spent 15 years teaching without a formal salary, funded by the students who attended his classes. He only obtained a salaried professorship in 1770 at the age of 46. He was a frail man just one and half metres tall, but many have noted that he was very social and frequently held lively dinner parties where his guests were encouraged to talk about anything but philosophy.

Kant was one of the most influential philosophers in the history of Western philosophy, particularly in ethics and metaphysics. His acclaimed published works on philosophy were written after he was 50 years old, when he was a professor of logic and metaphysics at the University of Königsberg. What I find most fascinating though, are his ideas in astronomy and cosmology that he developed in his early 30s.

In 1754 the Berlin Academy announced a prize to be given to the person who could answer the question of why Earth's rotation might be slowing down – a finding of Edmund Halley. Kant submitted an essay that he would also publish in his town's weekly magazine. He correctly argued that this could be due to the gravitational tidal force of our Moon. As a consequence, the length of the day should slowly increase. Kant went on to speculate that the rotation rate of the Earth should decrease until it always turns the same side towards the Moon, that is, until the length of the day increases to match the orbital timescale of the Moon. Kant was correct in his speculations. However, he did not receive the prize which was awarded to another scientist for an incorrect idea!

It was Immanuel Kant who provided the first cosmological model based on naturalistic views in his 1755 book, 'Allgemeine Naturgeschichte und Theorie des Himmels'. Kant's ideas were unique in Western cosmologies as he considered an evolving universe. He suggested that starting with a primeval chaos of particles distributed throughout an infinite void, the denser particles would begin to attract the more tenuous and form condensations. Given the laws of nature, this initial chaos must evolve into regular and orderly structures. In 1796 the brilliant French scientist Pierre-Simon Laplace extended the ideas of Kant to develop what would become the modern theory of planet and star formation, now known as the Kant-Laplace nebula theory[11].

Kant argued for an ever changing and evolving universe in which time flows on over billions of years during which new stars and worlds must be forming. He was also the first to remove the guiding hand of god. Kant put god's role as the creator after which god became a spectator - that the planets move not by the work of god, but by an inherent property of matter.

---

[11] I would love to describe progress in these areas, particularly over the past few decades. However, our progress is the result of the work of hundreds of my colleagues and it would be unfair to single out any one person.

Extending some writings of the English astronomer Thomas Wright, Kant speculated that the stars in our Milky Way formed within a rotating disk-like structure that would maintain its stability against collapsing. He speculated that the fuzzy nebula, which the telescopes of the time could not resolve, were very distant collections of stars like our Milky Way – their elliptical shapes occurring from viewing disks at an angle, an idea that was rejected by nearly all at the time but which was confirmed by Edwin Hubble over a century and a half later. His so called 'island universes' were themselves members of even larger structures, an idea that was confirmed by three-dimensional galaxy surveys in the 1980s. In Kant's universe, structures died *"devoured by the abyss of eternity"*, only for new cosmic structures to emerge. He speculated that the entire world might return to a chaotic state and then re-emerge, possibly an infinite number of times.

Kant was not the first by any means to consider a cyclical universe. The Rigveda also talks about the Brahmanda, an oscillating universe that expands from a single point to collapse once again. The concept of a cyclical universe is still discussed by cosmologists today, but in the framework of a multiverse – the idea that patches of the universe are constantly creating new universes such that a near infinite number of universes exist, each embedded within the old. We will hear more about this in the very last chapter.

It is interesting that several early cultures thought that the universe was cyclical – that the ashes of a dying universe gave rise to a new universe. The early Vedic cosmology, for example as detailed in the 2nd century BC Manusmṛiti, even equated cycles of the universe (the life of Brahma) with timescales comparable to modern values of its age. The later Hindu concepts of innumerable universes with innumerable worlds, detailed within the Bhagavata Purana, predates our modern ideas of the multiverse by over a thousand years.

Kant was prescient in his ideas that would be discovered to be correct over a century later. And we will meet many of the discoverers shortly. Some scientists believe that only the discoverer of evidence deserves credit and not the scientist who proposed the idea. And even then, sometimes the more famous scientist at the time is credited for the discovery that was made by another. In the case of Kant, his ideas began the path toward modern cosmology and he certainly deserves recognition along with those who proved his ideas to be correct.

Throughout his life Kant reflected on the nature of space and time. In his famous 1781 work 'The critique of pure reason', Kant reasons that time itself is not a property of the cosmos, rather time is a creation of the human mind. He uses the following antinomy: The world must have a beginning in time, otherwise an infinite amount of time – an 'eternity', as Kant called it – would have already passed in this world – but no infinite series can be completed. On the other hand, the world can't have had a beginning in time, because this would imply a period of empty time before the world came into being, and nothing can come into being in empty time, as there isn't anything to distinguish one moment in empty time from another. Later, we will learn more about those scientists who shed light on the nature of space and time and on the emergence of our universe

Kant's tombstone in Kaliningrad has a short text from his famous 1788 work 'Critique of practical reason' written in German: *"Two things fill the mind with ever new and increasing admiration and awe, the more often and steadily we reflect upon them: the starry heavens above me and the moral law within me."* In his published work the text continues: *"I do not seek or conjecture either of them as if they were veiled obscurities or extravagances beyond the horizon of my vision; I see them before me and connect them immediately with the consciousness of my existence."*

9. Henry Cavendish (1731-1810)

"for determining the density of the Earth and the absolute strength of gravity"

In the 17th century the density and mass of the Sun and other planets were known only in terms of Earth's density, not as absolute values. To determine the absolute values in terms of substances that could be measured, or a standard weight, it was necessary to measure the strength of gravity – the attractive force of a given amount of matter. This was achieved thanks to one of the first precision experiments carried out by the English scientist Henry Cavendish in 1798.

Cavendish was born into one of the wealthiest families in England and he was able to fund his own well-equipped laboratory for his experiments. Incredibly shy, he was terrified of women, dropped out of university because he was afraid of speaking in public and communicated with his servants only in writing. He discovered many new laws of physics, but these were only revealed in his unpublished notebooks long after his death. In a 1996 biography of Cavendish, he was described as *"one of the greatest scientists of his century, one of the richest men of the realm, a scion of one of the most powerful aristocratic families, a scientific fanatic, and a neurotic of the first order"*[12].

Amongst his numerous discoveries were the existence of different types of gases and he demonstrated that air was a composite mixture. He also discovered hydrogen, which would later be found to be the most abundant element in the universe. Cavendish collected the bubbles of gas that were produced when iron filings were added to sulphuric acid. He showed that the gas was different to other gases and produced water when it was burned. Cavendish had discovered the element hydrogen (its name means water-generating) whilst at the same time revealing the compound nature of water and

---

[12] "Cavendish", Jungnickel and McCormmach (1996), American Philosophical Society.

ending the widespread belief that it was a fundamental element.

Cavendish is most famous for his experiment to measure the density of the Earth. The first experiment he proposed was simple in principle, but difficult in practice. By hanging a weight from a line next to a large mountain, the gravitation of the mountain should pull on the weight changing the angle it makes relative to the fixed stars. The deflection angle is proportional to the density of the mountain and the density of the Earth. By estimating the density of the mountain, and knowing the size of the Earth, the Earth's mass and therefore its density could be determined. The experiment was carried out by other scientists since Cavendish did not like to leave his home and a deflection angle of about 1/1000th of a degree was determined. However, this was a difficult experiment to undertake and the errors were great. Also, the density of the mountain had to be estimated. Unhappy with this, Cavendish carried out his famous measurement in the basement of his home – the first precision experiment with a control over systematic errors.

The basic idea was to measure the attractive force between a large and a small lead sphere. The experiment used two small 2-inch spheres, each suspended from a wire that were connected to each other by a cross beam so they could rotate – a torsion balance. Each small sphere was suspended next to an 8-inch lead sphere such that their gravitational attraction would cause the torsion balance to turn slightly. Cavendish knew that any variations of temperature, magnetic field or disturbances of air, inside the room could destroy the accuracy of his measurements. He therefore set up the room so that he could observe the movement using a telescope from outside.

The accuracy he attained was remarkable. Cavendish found that the attractive force of gravity between the large spheres was tiny. The suspended smaller spheres moved just one tenth of an inch and from this he determined the density of our planet to be 5.48 times that of water – a value that is within one percent of today's measurements! In essence, Cavendish had measured the strength of gravity, or the so-called gravitational force constant.

Newton never wrote down his law of gravitational in an explicit form. He just stated it in words. At the time there was no standard system of units for length, mass and force – that would only come at the end of the 18th century. Newton's law of gravity has a constant, which we now call the gravitational constant. This describes the absolute strength of the gravitational force that a given mass has. Newton thought that this force would be too small to measure, even between a mountain and a test weight. By comparing the fall of an apple on Earth to the fall of the Moon towards Earth, the gravitational constant cancels out. Newton calculated the masses of the Sun and planets using his equations, but his numbers were in units of Earth masses (or more precisely, in units of the Earth's density).

To understand the history and future of the universe, we need to know the mass (and energy) of its contents. One of the main components of our universe are the stars. Knowing the gravitational force constant, the mass of the Sun could be accurately determined using Newton's equations of planetary motion which relate its mass to the distance from the Earth to the Sun, and the orbital period of the Earth.

At the time of Cavendish, the system of units did not include a unit for force – that was only introduced in 1873. The gravitational force constant could easily be determined from

Cavendish's results by inserting the size and mass of the Earth. Then we have everything we know to determine mass of our Sun. We find that our yearly motion around the Sun at a distance of 150 million kilometres requires the Sun to have a mass of 2.0×10^{30} kilograms. That is a little more than 1,000 times the mass of Jupiter and 330,000 times the mass of the Earth.

Today, our standard unit of mass is the kilogram – originally defined in 1795 as the mass of one litre of water, but shortly after that and all the way up until 2018 it was defined by an actual physical object – a cylinder of platinum kept in a vault in Paris that required three independent keys to open. Official copies were given to other countries to serve as their national standards. However, it was discovered that the standard kilogram was slowly getting more massive as it adsorbs molecules from the air. Consequently, after many discussions of alternatives, in 2019 the kilogram was defined in terms of the Planck constant (the constant that relates the energy of a photon to its frequency)[13].

Cavendish died in 1810, and long after this the Scottish physicist James Maxwell acquired his unpublished works. Many of the manuscripts detailed Cavendish's research on electricity from a century earlier. There was even an unpublished draft of a book on electricity. His experiments had revealed the inverse square law of electrostatic attraction, before the French physicist Charles-Augustin de Coulomb rediscovered it, and after whom the law is now named. Cavendish had also anticipated Ohm's law, that the electrical

---

[13] Planck's constant has dimensions of mass multiplied by the square of length and divided by time. By fixing Planck's constant (to the exact value $6.62607015 \times 10^{-34}$ m2 kg/s), the kilogram is then defined by our definitions of length and time.

potential across a conductor is proportional to the electrical current. At the time of Cavendish, there were no devices to measure such a current, so he used his own body, gripping the ends of electrodes with his hands and noting whether the shock reached his fingers, wrists or elbows. His experiment took place 45 years before those of the German Physicist Georg Ohm, after whom the law is now named. These and many other studies by Cavendish were carried out long before his death. I wonder if he was too shy or not confident enough to publish them, or perhaps he thought he had not performed the experiments to a high enough accuracy.

10. Friedrich Bessel (1784-1846)

"for measuring the distances to the stars"

When we talk about distances in cosmology, we use light years – the distance that light travels in one year. This is convenient since the typical distances between stars in our galaxy is a few light years. That's an enormous distance – it would take about 80,000 years to reach the closest stars to the Sun with our fastest rockets. So how were these vast distances first measured? The first person to succeed and measure the distance to a star other than the Sun, and to glimpse the scale of our galaxy, was the German astronomer Friedrich Bessel.

Soon after the invention of the telescope it was a race amongst astronomers to be the first to measure distances to the stars in the night sky. The race was not won until the 19th century, when the quality of telescopes and measuring devices had improved such that the parallax of nearby stars could be measured. The idea was simple – choose a star and make accurate measurements of its position relative to nearby stars over the course of a year. Because of Earth's motion around the Sun, if the chosen star were closer than its neighbours, its position in the sky should change due to the different viewing angles. Then simple trigonometry yields its distance. But because of the vast distances to the stars, this angle was a tiny fraction of a degree.

Bessel was born in the Westphalia region of Prussia. At the age of 14 he started a seven-year unpaid apprentice in a shipping company in Bremen. He taught himself navigation using the stars and soon became a valuable member of the company and began to receive a salary. This inspired him to spend all his spare time learning mathematics and astronomy. After educating himself with astronomy textbooks, he used measurements of the position of Halley's comet taken in 1607 to calculate its precise orbit. He presented his results to the astronomer Heinrich Olbers who recognised his talents and recommended that he be employed as an astronomer. Bessel gave up his career in shipping to become an assistant at a private observatory near Bremen. Despite not having a university education, his abilities as an astronomer became widely noted and at the age of 25 he was appointed as director of the Königsberg Observatory by King Frederick William III of Prussia in 1810. The following year Bessel was awarded a doctorate upon the recommendation of the famous German physicist Friedrich Gauss, which led to his position of professor of astronomy at Königsberg.

Bessel began to search for stellar parallaxes in 1818 but failed and did not try again for nearly two decades. Great discoveries in astronomy were often made by either new instruments and detection devices, or by large systematic surveys of the night sky. After hearing about potential results from other astronomers, Bessel came back to the problem armed with a 'heliometer' made by the German astronomer Joseph Fraunhofer. This is an optical device placed at the eyepiece of a telescope that can be used to accurately measure positions.

Bessel observed the star 61-Cygni over an eighteen-month period, measuring its position relative to neighbouring fainter stars. He wisely chose this star since he knew it had a high proper motion – its position slowly drifts across the sky due to its motion in space. He correctly assumed it must be close to the Sun. Over the course of six months, the Earth moves halfway around the Sun - this is the furthest apart that Bessel could put his telescope 'eyes'. Over the course of one year, the position of a nearby star will appear to move back and forwards relative to the distant stars.

He detected a tiny change in the angle of 61-Cygni relative to the fainter stars of 0.314 arcseconds (about one ten thousandth of a degree) with an error of 0.02 arcseconds. Knowing the size of Earth's orbit, he could then infer that the distance from the Earth to 61-Cygni was 658,000 times the distance from the Earth to the Sun, or about 10 light years away. This was within ten percent of the value we measure today. This observation not only revealed just how large our galaxy could be, but this was also direct observational evidence that the Earth orbited the Sun.

In 1829 the Royal Astronomical Society awarded Bessel its gold medal. John Herschel's words at the ceremony convey the significance of this result: *"I congratulate you and myself that we have lived to see the great and hitherto impassable barrier to our excursions into the sidereal universe that barrier against which we chafed so long and so vainly — almost simultaneously overleaped at three different points. It is the greatest and most glorious triumph which practical astronomy has ever witnessed...."*

The words also reveal that two other astronomers had simultaneously measured stellar parallax, yet Bessel was the first to publish results that were deemed accurate and later confirmed by other astronomers.

Prior to measuring the distances to the stars, astronomers had assumed that they were all equally bright, and therefore they must lie at different distances. The astronomers who were trying to measure parallax were often choosing the stars which appear brightest to the naked eye to observe thinking that the brighter stars must be the closest. But once the distances of the stars were measured, their true luminosities could be determined using the inverse square law by which the intensity of light falls with distance. It was then realised that stars come in a wide range of luminosities. The brightest stars to our eyes are often the closest, but they could also be an intrinsically faint star very nearby, or a very distant star that could be a thousand times as luminous as our Sun. Later, correlations between the brightness and colours of stars were discovered – fainter stars were red and brighter stars were blue. This in turn led to the theory behind the evolution of stars developed over a large part of the 20th century.

Bessel is best known amongst university students today for his mathematical 'Bessel functions', but he made many other discoveries. For thirty years Bessel rarely left his observatory and he was a very careful observer of the night sky. He created a catalogue of 75,000 stars visible in the Northern hemisphere, measuring their positions to an accuracy of one tenth of an arcsecond. His careful observations of the positions of stars revealed a periodic motion of the star Sirius, from which he deduced it must have a dark companion star that caused this motion. Later it would be confirmed that there was a faint companion – a white dwarf star that is the end state of stars like our Sun. In 1841 he used the varying swing of a pendulum to calculate that the Earth's shape was wider at the equator by one part in 299 – an accuracy that was not greatly superseded until satellite observations.

Bessel died in 1846 following a long mysterious disease that is now thought to have most likely been intestinal cancer. Two years after his death his work was immortalised in the famous prose poem 'Eureka' by Edgar Allan Poe: *"By dint, however, of wonderfully minute and cautious observations, continued, with novel instruments, for many laborious years, Bessel, not long ago deceased, has lately succeeded in determining the distance of six or seven stars; among others, that of the star numbered 61 in the constellation of the Swan. The distance ... is nearly 64 trillions of miles from us."*

Part II:
The scientists who discovered our place in time and space

Our understanding of the universe grew as advances in telescopes and technology allowed ever more precise views of the stars. By the beginning of the 19th century, religion was gradually displaced from universities and academic freedom was pioneered particularly in German universities. It is rare to find a reference to god in a scientific study after this time. Astronomers refined their telescopes and made more detailed observations; physicists developed their theories of light and matter.

Notable 19th century discoveries that I did not mention included the discovery of emission and absorption spectra of starlight by Joseph von Fraunhofer which led to the discovery of the composition of stars. The discovery of light beyond the visible spectrum by William Herschel who first detected infrared light by placing a thermometer beyond the red light in the spectrum of sunlight seen through a prism. Studies of energy and heat led to the laws of thermodynamics and the concept of entropy by Rudoph Clausius. That light appeared to be a wave was demonstrated by the famous double slit experiment of Thomas Young. There were also theoretical breakthroughs in our understanding of light and electromagnetism by James Maxwell.

By the end of the 19th century, astronomers had determined our place in the solar system and realised that our cosmos was a vast sea of stars extending thousands of light years in all directions. But what lay beyond? And how did all of this come to be? Let me now move into the 20th century and give credit to those unrewarded scientists who were eligible for a Nobel Prize but were overlooked or neglected. Unfortunately, many of these scientists have since died, but some are still alive and there is still time for them to receive recognition.

11. Henri Poincaré (1854-1912)

"for the principle of relativity, the discovery of chaos theory and for predicting the existence of gravitational waves"

You may have heard the name of the French mathematician and theoretical physicist Henri Poincaré in the media about a decade ago. This was when the Russian mathematician Grigori Perelman solved the 'Poincaré conjecture' and famously declined the prestigious Fields Medal[14] and the one-million-dollar Millennium Prize for its solution. Perelman cared nothing for fame or money and he was also upset at the mathematical community that tolerated famous colleagues who tried to take part of the credit. The conjecture was a statement about topology made by Poincaré in the year 1904. Over the following century many famous mathematicians attempted a proof but failed, until Perelman, who spent seven years in isolation solving the problem.

The English philosopher and mathematician Bertrand Russell described Poincaré as *"the most eminent scientific man of his generation – more eminent, one is tempted to think, than any man of science now living"*[15]. Poincaré was nominated 51 times for the Nobel Prize in physics, from scientists as distinguished as Nobel laureates Marie Curie and Hendrik Lorentz, but he was never awarded this honour. He died in 1912, the year Nils Dalen was awarded the physics prize *"for his invention of automatic regulators for use in conjunction with gas accumulators for illuminating lighthouses and buoys."* Dalen had received just one nomination from a member of the Swedish Academy of Sciences.

Many of Poincaré's nominators recognised his contributions to the theory of special relativity, but he also

---

[14] The Fields Medal is the mathematician's equivalent of the Nobel Prize, since mathematics was not listed by Alfred Nobel.

[15] Science and method, Henri Poincaré 1920, translated by Francis Maitland, preface by Bertrand Russell. Pubs. Thomas Nelson and Sons.

received nominations for several of his other profound discoveries. He was a true polymath who made contributions to science from mathematics and astronomy to fundamental physics and philosophy.

Poincaré was born into an influential family. His father was a respected professor of medicine at the University of Nancy and a pioneer of occupational medicine. His cousin was the President and Prime Minister of France. Poincaré's school teacher at the Lycée in Nancy, said that one day he would become a great mathematician. He went on to win first prizes in the concours général, a competition between the top pupils from all the Lycées across France. He then studied at the École Polytechnique in Paris, France's most elite university – he was the top qualifier in the entrance exams in 1873 and graduated just two years later.

During his university lectures Poincaré often appeared distracted, but this was because his eyesight was so poor that he learned through listening. His student notebooks are filled with hundreds of doodles as well as transcripts of the lecturers' words. Even as a student he became known for visualising complex mathematical problems, working out solutions in his head before writing them down. He would later become a professor of general astronomy at the École Polytechnique, whilst at the same time holding a professorship in mathematical astronomy and celestial dynamics at the University of Paris.

In 1910 the French psychiatrist Édouard Toulouse published a book about conversations he had with Poincaré in which he analysed his thought processes. He wrote that when attempting to solve a problem, Poincaré would start at the beginning, developing results from first principles, regardless of the progress made by others. Also, that Poincaré only undertook mathematical research from 10am until noon, and then again from 5-7pm. He would then read in the evenings so as not to be distracted during sleep. Toulouse explained, that Poincaré expected the crucial ideas in his research to come when he stopped concentrating on the problem!

Not all Poincaré's endeavours were successful. Upon joining the French Bureau de Longitudes in Paris in 1897, he tried to decimalise angles and time. Since the Sumerians and Babylonians, a circle was divided into 360 degrees. Poincaré proposed subdividing circles into 400 degrees such that a right angle was 100 degrees. He went further to propose an hour should contain 100 minutes and each minute would be 100 seconds. These ideas were soon rejected since all the existing maps and clocks would become worthless!

One of the great problems since the time of Isaac Newton is known as the three-body problem. Whereas Isaac Newton solved the motion of two gravitating masses, such as a planet orbiting a star, when it came to trying to understand the motion of three bodies – for example, the Sun, Earth and the Moon – a general solution eluded him. The problem is very well defined: given the initial positions and velocities of three objects interacting via gravity, how do their motions proceed over time? Over the next three centuries many famous scientists attempted to find a solution to this problem and failed. Poincaré showed why they failed and discovered chaos theory in the process.

The three-body problem was deemed so important that the King of Sweden, Oscar II, announced a prize in 1887, in honour of his 60th birthday, for anyone who could make progress on its solution. Poincaré's successful submission for the prize *'Sur le problème des trois corps et les 'equations de la dynamique'*, was 270 pages long. He showed that it is not possible to write down a general solution. Over his career Poincaré kept returning to the three-body problem and wrote three more volumes totalling 1300 pages on the topic. These works have since stimulated a vast number of studies in widely different research areas.

What Poincaré essentially showed was that the future behaviour of the three bodies was very sensitive to small changes in the initial conditions. This is the essence of chaos theory. The future and past motions of three bodies can in principle be determined from knowledge of their positions and velocities at any one instance in time – this is called a deterministic problem. In this case the solution could in principle be calculated but not even a computer could ever be powerful enough to calculate it exactly!

Poincaré went on to discuss weather predictability, which was largely forgotten until the 1960s when meteorologist Edward Lorenz rediscovered chaos theory when trying to make weather predictions with a computer. You may have heard of the 'butterfly effect', that the flap of a butterfly's wings can trigger a tornado far away. This was the title of a talk given by Lorenz in 1972 where he presented his results on the unpredictability of the weather. Lorenz found by accident that his weather predictions had completely changed when repeating the same calculation because of how his computer was rounding the last digit. In his 1908 book *'Science et méthode'*, Poincaré mentions the apparent random occurrence of storms and rain and writes *"Here again we find the same contrast between a minimal cause, too small to be seen by an observer, and substantial effects, which are sometimes terrible disasters."* Still today, the discovery of chaos theory is often attributed to Lorenz rather than Poincaré.

Poincaré was also fascinated by the geometry of space and the nature of time. He made major contributions to the development of the theory of relativity, a theory that played a key role in our discovery of the universe and in the development of physics in general. The theory of special relativity describes the connection between space and time. It is a set of equations that relate how things appear in one frame of reference to how they look in another. It is called special because it does not include gravity. It leads to weird but proven phenomena, such as time slowing down for astronauts, and it revealed the most famous law in physics, $E=Mc^2$.

The theory of special relativity is usually credited to Albert Einstein, yet historians of science still today try to piece together how Einstein came up with his famous 1905 paper, which has no references to other works. There are many books and research articles on the history of relativity, and I cannot do the topic justice here. But let me give the context of Poincaré's contributions, since I agree with those historians of science who argue that he deserves at least half the credit.

Since the work of Newton, and until the beginning of the 20th century, it was largely accepted that space must be filled with some form of medium that propagated light, heat, electricity and magnetism. This medium was called the ether and there was a different ether for each of these phenomena. Then, in 1865, the Scottish scientist James Clerk Maxwell showed mathematically that light itself was an electromagnetic wave and that such waves must always propagate through space at a constant speed - the speed of light. But if electromagnetic waves propagated through a medium then their speed should vary according to the motion of that medium. If not, then observers moving at different speeds would measure different laws of physics. Maxwell proposed ways to test the existence of the ether which culminated in the famous experiments of Albert Michelson and Edward Morley.

To detect the ether Michelson and Morley measured the speed of light in two directions – one parallel to Earth's motion and one perpendicular. It was thought that the speed of light should be faster in Earth's direction of motion as you would add its speed of motion to the speed of light. In a sense this is rather like what we experience when our voice travels further downwind, or when we throw something from a moving vehicle. But the universe constantly violates our expectations - the remarkable finding was that the speed of light was the same in both directions. No matter how fast you are moving, if you shine a beam of light, it will always travel at the same speed, no more and no less. This is the most famous null experiment ever performed since it did not reveal the expected presence of an ether.

Many scientists tried to make sense of all of this, including the Dutch physicist Hendrik Lorentz who in 1892 began to develop transformations between space and time of observers moving relative to each other through a stationary immobile ether[16]. The aim was to rewrite Maxwell's equations in a form that remained unchanged relative to a moving observer. However, Lorentz's calculations were incorrect. It was Poincaré who corrected the errors and also incorporated gravitation with the theory which he called 'the new mechanics'. He wrote down the correct transformations and named them for the first time 'Lorentz transformations'. He also showed that gravitational waves, disturbances that

---

[16] Unknown to Lorentz, the same transformations had also been written down by the German physicist Woldemar Voigt in 1887.

propagate through space, would be a natural consequence of his theories – a phenomenon that would only be discovered much later and for which a Nobel Prize was awarded in 2017.

That time is an absolute quantity which could be measured and agreed upon by all, was also universally accepted until to the beginning of the 20th century. It is a natural consequence of Newton's laws of motion. Poincaré questioned the existence of absolute time and came up with his 'postulate of relativity' before Einstein – that the laws of physics should be the same for all observers and that an observer could not discern whether they themselves were moving or stationary. He described how the clocks of two observers would only measure the same time if they were at rest. He wrote that the speed of light was the maximum speed that could exist and that gravity should propagate at the same speed otherwise his postulate of relativity would be violated. That gravity travels at light speed was verified in 2017. In July 1905 he submitted a summary of his ideas in a paper called the 'Memoir'. But because the journal to which he submitted to only published twice a year, his work did not appear until 1906. In the meantime, Albert Einstein published his famous paper on relativity in 1905.

In their separate papers on electrodynamics, Poincaré and Einstein explained that with the Lorentz transformations the relativity principle holds perfectly. Einstein also began with the postulates that the laws of physics should be the same for all observers and that the speed of light was a constant and independent of the motion of its source. From these postulates he derived the Lorentz transformations. Poincaré was looking for a physical reason as to why the speed of light appeared a constant in all reference frames, but Einstein took it as a property of nature. In the work of Poincaré, space and time were separate entities, whereas Einstein showed they were woven together.

Poincaré came awfully close to the complete theory of special relativity, but Einstein's treatment was simpler and self-consistent and did not require the existence of an ether. In Einstein's 1905 publication he never referenced Poincaré. However, a personal friend of Einstein, Maurice Solovine, acknowledged that he and Einstein pored over Poincaré's 1902 book 'Science and hypothesis', keeping them "*breathless for weeks on end*". This book was written for a non-specialist in which Poincaré describes his insights into unsolved problems, including the photo-electric effect, Brownian motion and the relativity of physical laws in space – problems that Einstein would resolve in his four famous papers from 1905.

Poincaré was the only scientist who became an elected member to each of the five separate divisions of the French Académie des Sciences; geometry, mechanics, physics, geography and navigation. He received several prizes and honours for his research, but not the Nobel Prize despite numerous nominations. He died aged just 58 following complications from surgery. Princes and presidents attended his funeral, during which he was addressed as a 'poet of the infinite' and a 'bard of science'.

12. Albert Einstein (1879-1955)

"for the theory of general relativity that connects space, time, matter and energy"

Albert Einstein is undoubtedly the most famous scientist of all time. And he deserves that credit twice over. His most famous work is without question his theory of general relativity. Without this theory, we would be unable to understand the history and future of our universe, and far more. You may be surprised to learn that Einstein was never awarded a Nobel Prize for this work.

Albert Einstein was born in Ulm, part of the Kingdom of Württemberg in the German Empire in 1879. He recalls at the age of five being fascinated by the mysterious invisible forces that moved the needle of a compass. Aged 16 he pondered the question, what would a beam of light look like if you could pursue it? If light were a wave and you travelled alongside it, then surely it should appear stationary, he thought, but stationary light waves had never been observed. This thought experiment played a role in many of his later discoveries.

In 1895 Einstein moved to Switzerland to study after disliking the education he was receiving in Munich, as well as having a dislike of the forced subscription to the military. Contrary to folklore, he did well at math, mastering differential and integral calculus by the age of 15. However, he failed the entrance exam to the Eidgenoessische Polytechnische Schule (now the ETH Zurich), which he attempted two years younger than most students. He went back to high-school in Aarau to improve his knowledge and to obtain his Matura which would allow him to study at university. After he gained his diploma in 1900 from the Eidgenoessische Polytechnische Schule, he acquired Swiss citizenship. Even though he had given up German citizenship partly to avoid military service, he reported for military training as a Swiss citizen in 1901. According to his medical examination, he suffered from varicose veins, sweaty and flat feet, and he was happy to be rejected from military service!

Despite applying to many universities for a teaching position, he failed to receive an offer. He found employment at the Swiss patent office in 1902, and in his spare time wrote his PhD thesis on 'A new determination of molecular dimensions' (Eine neue Bestimmung der Moleküldimensionen). For this he received his doctorate from the University of Zurich in 1905.

In the same year, whilst still working at the patent office, Einstein published four papers in the Annalen der Physik, and because each of them were major advances in physics, they are called the Annus Mirabilis papers (from the Latin 'miracle year'). The first paper explained the photoelectric effect, that when you shine a light on a material you can observe the emission of electrons. This could not be explained by classical physics at the time since light was treated as a wave propagating through space, but Einstein showed that it could be explained if light were a stream of particles (called photons) that could only have particular energies. This was a crucial step in the development of the theory of quantum mechanics. The second paper explained Brownian motion, which is the random motions of particles suspended in a liquid. It was first noted by the botanist Robert Brown who through a microscope saw pollen jiggling around on the surface of water. Einstein showed that this was due to the random motions of individual water molecules colliding with the pollen. This was the first convincing evidence of the existence of atoms and molecules.

Einstein's third paper was the theory of special relativity that we learned about in the last chapter. And his final paper of the year was about the relationship between mass and energy – that mass can be considered a form of energy (or energy can be considered a form of mass) related by a simple proportionality constant (the square of the speed of light); $E=Mc^2$, the most famous equation in science.

Einstein spent seven years at the Swiss Patent Office, but by 1909 his research was becoming widely recognised and he received a professorship in theoretical physics at the University of Zurich [17]. Whilst in Zurich Einstein began thinking about a more general theory of relativity that incorporated gravity and the new mathematical description of spacetime.

Up until 1907 space and time were still considered as separate entities. It was the German scientist Herman Minkowski who showed that relativity could be understood very neatly if it were described in a four-dimensional space, called spacetime. In his 1908 address to the 80th Assembly of German Scientists, Minkowski dramatically began with the words: *"The views of space and time which I wish to lay before you have sprung from the soil of experimental physics, and therein lies their strength. They are radical. Henceforth space by itself, and time by itself, are doomed to fade away into mere shadows, and only a kind of union of the two will preserve an independent reality."*

That is easy to say, that time is another dimension of space that we call spacetime, but it is not something we usually experience. We are familiar with the relentless passage of time. We are familiar with moving in space. But if we move through space then we also appear to move through time at a different rate, relative to someone watching us move. It is a tiny

---

[17] When I joined that same institute as professor for theoretical physics in 2002, my first request was to see Einstein's original PhD thesis. It is the shortest thesis I have ever held, just 17 pages long, but 17 remarkable pages.

difference, because our fastest transport is much slower than the speed of light, but it is a measurable effect.

By 1915 Einstein had completed his major work on the geometric theory of gravitation which connected gravitation, energy and spacetime. It is the framework by which we can understand the universe and its contents, from the development of the big bang to the workings of black holes.

Einstein's path to the derivation of general relativity can be seen in his famous 'Zurich Notebook', which contains his handwritten notes from 1915. The notebook begins from both the front and the back simultaneously; his calculations meet in the middle. The first page from the back has the sketch and designs for a complex children's puzzle. Then he delves into electrodynamics and relativity, as well as the mathematics needed to describe the structure of spacetime. This is the beginning of further notes and equations in which he develops the formidable theory of general relativity – a unified theory of gravity that links together the geometry of spacetime with the matter and energy that it contains.

In Einstein's theory of gravitation, objects move not because of an instantaneous force acting between matter, but because objects travel along paths in space that is curved by the presence of matter and energy. Newton's laws follow from general relativity only if you assume that the speed of gravity is instantaneous. But just as Poincaré had, Einstein postulated that gravity, like light, travels at a finite and constant speed in any reference frame.

Einstein used general relativity to predict the deflection of light by massive objects (a consequence of gravity curving space), the presence of black holes (places where mass is so concentrated that spacetime curves back on itself), that light would be redshifted by gravity (a consequence of gravity affecting time), the existence of gravitational waves (a consequence of gravity propagating at the speed of light) and much more – and every prediction has so far been proven correct. Most importantly for our understanding of the history and future of the universe, matter and energy (they are both equivalent) affect space and time. If you know the matter and energy content of the universe, you can calculate exactly how it should evolve in time!

In 1917 Einstein applied his equations to the universe as a whole – a static and unchanging universe - and he found the same problem as Newton, that it was unstable and would collapse. Einstein attempted to create a stable universe by adding a new constant to his equations which provided an opposing positive energy to the negative energy of gravity – he called it the cosmological constant. This term countered the gravitational attraction and prevented his model universe from collapsing. He later regretted not considering dynamical universes, universes with an origin in time.

Einstein's theories of relativity changed the entire field of physics and cosmology. In 1919 the British astronomer Arthur Eddington observed the deflection of starlight around the Sun, confirming one of Einstein's predictions of relativity. Einstein became a scientific celebrity overnight. Earlier that same year, he had divorced from his first wife, Mileva Maric, and despite his growing fame he was not wealthy. In his divorce settlement he pledged that if he received the Nobel Prize then Maric and their children would receive the prize money. But despite being nominated 62 times for a Nobel Prize for his theories, by 1921 he had still not been awarded the honour.

In 1921, the Nobel Prize committee received 14 separate nominations from famous scientists recommending that Einstein should receive the prize for his theory of relativity.

The Nobel Committee commissioned one of their members, Allvar Gullstrand, professor of ophthalmology at the University of Uppsala, to prepare an account of the theory of relativity. Gullstrand was one of five members on the panel which decided on whom to award the Nobel Prize in physics[18]. Gullstrand's report was highly critical of Einstein's work stating that the effects of relativity are so small that they lie below the limits of experimental error. This was incorrect as relativity had been confirmed by two separate experiments. He also wrote that Einstein's theory was *"a matter of unproven belief and not of greatest utility to mankind"*. The Nobel committee decided to not even award the prize in physics that year.

By 1922 it was getting embarrassing for the committee as they received 17 new nominations for Einstein. One of those nominations was for his work on the photoelectric effect by the Swedish physicist Carl Wilhelm Oseen, who had also nominated Einstein in 1921. The Nobel committee again commissioned Gullstrand to provide an updated report on relativity, whilst Oseen was asked to report on the photoelectric effect. Gullstrand's report was still highly critical, whilst Oseen's report was, unsurprisingly, very positive. They decided to award the unawarded 1921 prize in retrospect to Einstein for *"services to theoretical physics, and especially for his discovery of the law of the photoelectric effect,*

---

[18] Gullstrand is the only person to have both declined and received a Nobel Prize. He was awarded the prize in physics in 1911 for his work on the optics of the human eye, whilst a member of the physics committee. However, he declined the award because he wished to be awarded the prize in medicine which he was also awarded in 1911. This was all done rather discreetly and the 1911 prize for physics was then awarded to Wilhelm Wein.

without taking into account the value that will be accorded your relativity and gravitation theories after these are confirmed in the future".

This is the only time that the Nobel Committee included a note about what the prize was not awarded for! Einstein took this as a slap in the face as his theory of relativity had been confirmed twice and he, like many other scientists, thought it was his greatest work. Since Einstein was touring Japan, he did not collect his prize in person, and the German ambassador accepted the prize on his behalf, stating this was a great day for German science. Einstein was also upset by this – he and his daughter pointed out that he was actually a Swiss citizen. Later, when Einstein was asked about which of his many prizes had made him most proud, he listed many but did not even mention the Nobel Prize, despite it having the highest profile and largest pay out. The best to come out of all of this was that Maric eventually received the prize money, 121,572 Swedish Kronor, equivalent to 12 years of Einstein's salary at the time.

Einstein moved to Germany in 1914 just before the start of World War I, attracted by the offer to head a prestigious new institute for physics in Berlin, and to be close to his cousin Elsa Loewenthal whom he would marry in 1919. In 1914 Einstein made his first openly political statement by condemning German aggression in World War I; he was one of only four German scholars to sign the Manifesto to Europeans to protest Germany's aggression. Throughout the rest of his life he would use his fame to promote pacifism.

Einstein and Elsa left Germany for America in 1933 to escape the sudden rise of Hitler to power. In 1939 Einstein heard from colleagues that Hitler was undergoing research into the construction of an atomic bomb. He wrote a letter to

President Roosevelt recommending that the United States pay attention and develop its own nuclear weapons research. Many think that this initiated the Manhattan Project – an Apollo scale project to produce the first nuclear weapons. In 1954 Einstein told his colleague, Linus Pauling, that this was the one great mistake in his life. In 1955, a few days before his death, Einstein signed the 'Russell-Einstein' manifesto calling attention to the horrors of a third world war, and an end to international conflict and the use of nuclear weapons.

You will see more 'quotes' by Albert Einstein reproduced in books or the internet, than any other person. But many of them are incorrectly attributed. Another quote that is widely attributed to Einstein may also be not his words, but I like it and can imagine him saying: *"I do not know with what weapons World War III will be fought, but World War IV will be fought with sticks and stones."*

13. Henrietta Swan Leavitt (1868-1921)

"for the discovery of an astronomical standard candle that led to the determination of our place in the Milky Way galaxy and within the universe"

Measuring the distance to our Sun was the first step in the cosmic distance ladder which was necessary for our discovery of the universe. We heard how this was accomplished in chapter 7 on Edmund Halley. The second step was the distance to the nearby stars using parallax first measured by Friedrich Bessel which was described in chapter 10. Because of the limitations of the telescopes of the time, parallax could only be used to measure the distances to the closest stars. The true extent of our galaxy was not realised until 1908 when the American astronomer Henrietta Swan Leavitt discovered a way of measuring the distances to much more remote stars, revealing that our galaxy was at least 50,000 light years across. Leavitt discovered what we now call, 'a standard candle'. This technique is the third step in our distance ladder which enabled astronomers to determine the structure of our galaxy and the discovery that our galaxy is just one of countless others in the universe. It also enabled the discovery of the big bang.

Imagine a light source that is always the same strength. Astronomers call these 'standard candles' and use them to measure distances to faraway objects. For example, if all stars had the same brightness, more distant stars would always appear fainter. This is because the number of photons we receive diminishes as one over the square of its distance, just like the force of gravity. A star that was four times fainter than another would by implication be twice as far away. In this case, knowing the distance to one star would allow us to estimate the distance to any star.

In 1698, the Dutch astronomer and physicist Christiaan Huygens estimated the distance to the star Sirius by comparing its brightness with that of the Sun. It was hard to measure the equivalent brightness of the Sun. But by using a pinhole to reduce the amount of sunlight until it appeared as bright as Sirius, Huygens estimated that Sirius must be 27,000 times as distant as the Earth–Sun distance. Since Sirius is the brightest star, it was thought it must be the closest. However, the assumption that all stars have the same brightness (that they are all standard candles) was incorrect. At this time, it was not known that stars have a vast range in their brightness (luminosities). In fact, Sirius is much brighter than our Sun – about 25 times brighter – so Huygens' distance estimate was short by about a factor of five.

Henrietta Leavitt discovered an accurate stellar standard candle whose brightness and distance could be inferred from a simple observation.

Leavitt's father was a church minister and her family was not super rich, but wealthy enough that she could attend Harvard University's Society for the Collegiate Instruction of Women. At the time, Harvard did not admit women but they were allowed to attend lectures. In her fourth year of studies she attended a course in astronomy which greatly inspired her. In 1895 she became a volunteer assistant at the Harvard Observatory.

Photography became a powerful tool in astronomy towards the end of the 19th century and Harvard Observatory was systematically photographing the night sky. This led to hundreds of photographic plates each with numerous stars. In the early 1900s women were not allowed to operate telescopes, but Leavitt and a few dozen other women were hired by Edward Pickering, director of the Harvard University Observatory, to perform the rather dull task of measuring and cataloguing the brightness of thousands of stars on the photographic plates.

These women were known as the 'computers'. Initially Leavitt was not paid for her work but later received 30 cents per hour. This is often quoted as a low salary but was rather typical of the average American worker at the time. She carried out her duties despite being almost completely deaf after an illness. Her task was to identify variable stars, whose brightness appeared to change. She identified nearly 2000 such stars. In 1912 she made her famous discovery. When she compared the brightness of the same stars on images taken on different nights, she noticed that they varied in a well-defined and periodic way – the luminosity of the brighter stars was changing over a longer period of time than the fainter stars.

The stars she was studying are called 'cepheid variables'; they pulsate on a timescale from one day to a couple of months, leading to a regular change in their brightness. Their regular pulsation is a consequence of the escaping radiation that causes the outer regions of the star to expand. As it expands, it cools down and contracts, and the process begins again. Leavitt noticed that any cepheid variable star whose visual brightness changes with the same period will have the same total luminosity. Therefore, just by measuring the pulsation period and apparent brightness of a cepheid variable star, we can determine its distance. This technique relies on knowing the distance to at least one nearby cepheid variable, which as Leavitt pointed out, could be measured using the technique of parallax. Pickering published her results as his own, although the first sentence of the publication states *"prepared by Miss Leavitt"*

You can witness the regular change in some cepheid variable stars by eye. A good example is Delta Cephei whose brightness changes by over a factor of two every five days. It lies in the constellation of Cepheus, next to two stars that are conveniently as bright and as faint as Delta Cephei varies, so you can tell how bright it is on a given night. It is due to this star that cepheid variable stars obtained their name.

Cepheid variable stars are very luminous and can easily be identified in the distant regions of our galaxy and even in nearby galaxies. The astronomer Harlow Shapley learned of Leavitt's results and had access to the world's largest telescope at the time – the 60-inch reflector at Mount Wilson in Pasadena. He measured the parallax to several of Leavitt's variable stars to determine their distances and to calibrate the period-luminosity relation. By 1918 he had observed variable stars in the remote regions of the Milky Way, determined its vast size and discovered that our Sun was not at its centre. Leavitt's fundamental contribution to this work was barely mentioned by Shapley.

At this time, the known universe consisted only of the stars within our Milky Way galaxy. Nothing else was known to lie beyond it. But a debate had been brewing about the nature of fuzzy patches of light amongst the stars that could be seen by eye. The first mention of these strange objects was made in 964 by the Persian astronomer Abd al-Rahman al-Sufi, who noticed a faint extended patch of light where we now know the Andromeda galaxy to be located. The so-called nebulae were noted to be extended blurry patches of light, neither starlike nor cometary in origin. Were those fuzzy patches of light clouds of gas and stars in our own galaxy, or systems like our own galaxy but outside and far away? Earlier we learned that the first to speculate that the nebulae were 'island universes' beyond our own galaxy, was the 18th century philosopher Immanuel Kant.

For hundreds of years the nature of the fuzzy nebulae was unknown and much discussed, until the American astronomer

Edwin Hubble used the technique pioneered by Leavitt to measure the distances to the Cepheid variable stars that lie within the closest of these nebulae – the Andromeda galaxy. Hubble had access to the new one-hundred-inch telescope at Mt Palomar in California. He measured the pulsation period and apparent brightness of Cepheid variable stars within the Andromeda nebula. Using this next step in the distance ladder, in 1925 Hubble found that the Andromeda nebula was at least 10 times further away than the most distant stars in our galaxy.

Andromeda is the closest large galaxy to our Milky Way and lies over two and a half million light years away. And beyond that are countless others. It was another profound realisation that the universe was a truly vast place containing numerous other 'island universes' – galaxies like our own each containing hundreds of billions of stars just as Kant had speculated.

In 1925 the Swedish mathematician Gösta Mittag-Leffler wrote Leavitt a letter: *'Honoured Miss Leavitt, your admirable discovery ... has impressed me so deeply that I feel seriously inclined to nominate you to the Nobel Prize in Physics for 1926'*. He was unaware that Leavitt had died from cancer four years previously. Leavitt certainly deserved a Nobel Prize since her discovery was fundamental to our understanding of our universe. Moreover, it was essential in discovering the big bang. Today, most astronomers still name her discovery as the 'cepheid period luminosity relation'. It should surely be called 'Leavitt's law'.

Henrietta Leavitt died aged 53 from stomach cancer. Solon Bailey was her senior colleague at the Harvard University Observatory and wrote the only obituary of her that I could find. He wrote *"She took life seriously. Her sense of duty, justice and loyalty was strong. For light amusements she appeared to care little. ... She had the happy faculty of appreciating all that was worthy and lovable in others, and was possessed of a nature so full of sunshine that to her all of life became beautiful and full of meaning.[19]"*

---

[19] Popular Astronomy, 1922. S. Bailey, Vol. 30. P. 197.

14. Georges Lemaître (1894-1966)

"for the discovery that the universe is expanding and had a hot and dense beginning – the big bang"

By 1925 astronomers had revealed that not only was our Sun insignificant within our vast galaxy of stars, but that our galaxy was just one of numerous others within a vast universe. The next breakthrough, and certainly one of the biggest findings in the history of science, was the discovery that our universe is expanding, which implies that it had a beginning. The discovery of the big bang was made by the Belgium priest and scientist Georges Lemaître.

Georges Lemaître was born in Charleroi in 1894 in a well-off, devoutly Catholic family. He began to think about the universe at an early age: while in the trenches during World War I he wrote to his friend *"I have understood the 'Fiat Lux' [Latin for 'let there be light'] as the reason of the universe"*. During the following years he was simultaneously studying to become a priest and also studying Einstein's recently published works on relativity. In 1923 he moved to Cambridge to study under the astronomer Arthur Eddington and began to think more deeply about the origin of our universe. Eddington wrote *"I found Lemaître a very brilliant student, wonderfully quick and clear-sighted, and of great mathematical ability."*

At the same time as astronomers were still developing the tools and techniques needed to observe and measure the universe, theoretical cosmologists were busy trying to find solutions to Einstein's equations so they could apply them to the universe. They started with the assumption that space is isotropic (similar in all directions) and that the universe is homogeneous (similar in all places). This is called the Copernican principle that I mentioned in the chapter about Copernicus as being central to modern cosmology.

In 1922 the Soviet physicist Alexander Friedmann explored solutions to Einstein's equations in which the universe was not static, but dynamic. Friedmann discarded the cosmological constant and explored the behaviour of an expanding universe. He calculated models in which the universe could slow down its expansion and collapse, or how the universe might keep on growing in size forever. He showed how these future behaviours depended on the matter content of the universe. At the time, there was no evidence for a dynamical universe and there were no measurements of its mass. But he speculated its mass could be equivalent to 10^{21} stars like our Sun, in which case it would be around 10 billion years old. It was an amazing feat of human intellect to be able to contemplate such radical ideas, and to think about models in which the universe had a finite age. Unfortunately, Friedmann died in 1925 a few years before the big bang was discovered.

The discovery of the big bang was enabled by the American astronomer Vesto Slipher who had measured the motions of nearby galaxies using the doppler effect. If a light source is moving away from us, then the wavelengths of all the photons that it emits will be stretched by an amount proportional to its speed. In 1917 he reported measurements for 25 such nebulae and 21 of them were moving away from the Sun, one of them at 1800 kilometres per second!

In 1925 Georges Lemaître became associate professor of mathematics at the Catholic University of Louvain, and by 1927 he had put all of these findings together. He calculated solutions to Einstein's equations for an expanding universe and associated the expansion, not with a physical velocity of galaxies, but with an expansion of space between the galaxies. He then took the velocities measured by Slipher and the distances measured by Hubble, and showed that there was a linear relation between them. Galaxies further away appeared to be moving away from us more quickly. This is exactly as expected if the universe were expanding – as the space between the galaxies increases, the wavelength of light travelling across this space is stretched. Lemaître had discovered the big bang[20].

Prior to this discovery it seemed natural to conclude that the universe did not change over very long timescales. The incessant motion of the Earth around the Sun, which appeared to be the same each day, and the stars that appeared in the same place each night, all gave the impression that the universe was static and unchanging. Lemaître's results completely changed our perception of the universe. It was not a static timeless entity. Rather, it was rapidly expanding, which implied that it was smaller in the past and ultimately originated from a small and very dense region – that it had a beginning.

But that was not the end of the story. Most scientists at the time who heard of Lemaître's work disliked the idea of an expanding universe. Upon meeting Lemaître in 1927, Einstein is reported to have told him that his math was correct, but his physics was abominable. Einstein clearly disliked the concept of an expanding or contracting universe, in particular the initial singularity it implied, but he soon began exploring its consequences in more detail. Then, in 1930, Eddington re-examined Einstein's static universe model and showed it was not stable. It was as unstable as balancing a pencil on its tip. Eddington realised that Lemaître's expanding universe solved this problem. He sponsored the translation of Lemaître's work into English to raise awareness of just how profound it could be.

---

[20] See Appendix I for an overview of the big bang.

Lemaître had originally published his work in a French scientific journal, which is often credited as the reason why Edwin Hubble is generally credited for this discovery. I rather think it is simply the fact that Hubble was already a famous astronomer at the time. In 1929 Hubble published a paper titled, 'A Relation between Distance and Radial Velocity among Extra-Galactic Nebulae'.

He simply plotted the data for the galaxies and made no link to a possible expanding universe. Hubble took the velocities as real motions of the galaxies flying apart, rather than an expansion of space. He never really believed in the ideas behind the big bang. In 1929 Hubble wrote *"It is difficult to believe that the velocities are real; that all matter is actually scattering away from our region of space. It is easier to suppose that the light-waves are lengthened and the lines of the spectra are shifted to the red, as though the objects were receding, by some property of space or by forces acting on the light during its long journey to the Earth."*

In 1931 Lemaître went one step further. He published a short text in which he introduced the concept of the 'primeval atom', explaining that if the universe were expanding today, it must have once been condensed into a tiny space, comparable to an atom that contained a vast amount of energy. In his theoretical models he re-introduced Einstein's cosmological constant to create a universe that was old enough to match the age of the Earth. Despite the evidence, his profound discovery was largely unaccepted - even Eddington disliked the notion of a beginning of time and wrote *"Philosophically, the notion of a beginning of the present order of Nature is repugnant to me"*[21].

---

[21] 'The End of the World: from the Standpoint of Mathematical Physics', 1931, Nature, No. 3203,

That the universe is expanding put to rest a paradox that was first raised in the 6th century by Cosmas the Monk, a Greek explorer. The puzzle is that if the universe was infinite in size and hosted an infinite number of stars, then why is it dark at night? This is now known as Olber's paradox, after the German astronomer Heinrich Olber who described the problem in 1823 even though many before him, including Cosmas the Monk and Johannes Kepler had also posed the same question. In an infinite universe of stars, every line of sight should end on the surface of a star so the sky should be as bright as the surface of the Sun. It is rather like standing in a large forest and in every direction you see a tree. Cosmas argued that the crystalline shells that moved the planets and hosted the immovable stars must also absorb their light. It was the poet Edgar Allan Poe who came close to the correct explanation in his poem 'Eureka'. That the universe is finite in age and because light travels at a finite speed the light from distant stars has not yet reached us. This was partly correct, as we have learned that the expanding universe must have had a beginning in time. The fact that the universe is expanding provides the full solution to the paradox, since the light from the most distant galaxies is stretched into longer less energetic wavelengths that we cannot see.

Lemaître was the scientific 'father' of the big bang although his contributions became largely forgotten. At the same time, he was a devout christian who rejected the abuse or use of his own science to defend religious doctrines. He separated religion from science with the view that the two should not be compared, that the bible does not teach science. He disliked pope Pius XII's use of his ideas to defend the doctrine of creation by god, stating that *"the pursuit of truth is the highest human activity"*.

Today, there is no observational evidence that does not support the big bang theory. The discovery of Lemaître ranks alongside the greatest of all discoveries. Despite this, most people, even most professional astronomers, associate the discovery of the big bang with Edwin Hubble. Hubble has received numerous accolades and honours for his 'non discovery'. NASA even named the Hubble Space Telescope after 'the discoverer of the big bang'. I suppose that naming it after the Belgian scientist Lemaître who actually discovered the big bang was not an option. Lemaître was nominated just once for the Nobel Prize in 1953.

15. Arthur Eddington (1882-1944)

"for the energy source and the first theoretical models of stars"

Arthur Eddington was the world's most famous astrophysicist during the early 20th century. We have already come across his name in earlier chapters. Eddington was born in Cumbria, England in 1882 and was brought up by his mother since his father died of typhoid when he was just two years old. His interest in stars began at a young age. When he was given a small telescope, he tried to count the stars in the night sky and he set himself the task of counting the letters in the bible[22]. He excelled at school and earned a scholarship to go to Manchester university at age 15, and later another scholarship to study at the University of Cambridge. There he became the first 2nd year student to gain the title of 'Senior Wrangler', the top mathematics student. Already by 1906 he had become the chief assistant to the astronomer royal at the Royal Greenwich Observatory and within a decade he directed the entire Cambridge Observatory.

During the horrors that were occurring during the First World War, many British scientists were arguing that scientific relations with Germany and Austria should be permanently ended. Eddington, a vocal pacifist, was virtually alone in his arguments to keep politics out of science. He was the only astronomer who tried to maintain relations with his European colleagues and was the first to learn of developments by a young Albert Einstein. Eddington was fortunate in being one of a handful of astronomers with the mathematical skills to understand general relativity and one of the few who would have been interested in pursuing a theory developed by a German physicist. And because Einstein was also a pacifist,

---

[22] Allie Vibert Douglas: *The Life of Arthur Stanley Eddington*. London 1957: 5

Eddington was happy to promote his work to spread a message of peace.

Eddington was a shy, withdrawn man, a devout Quaker, and a strict teetotaller. He remained a bachelor all his life. As a student he kept a diary in which he wrote about his travels and vacations, but in later years he only wrote down the sales figures from his books and the distances he cycled - he was a passionate bike rider, often riding over 100 kilometres a day.

During the war Eddington obtained an exemption from compulsory conscription to the army because of his forthcoming expedition to observe the solar eclipse in 1919 on the island of Principe off the West coast of Africa. According to Einstein's theory, the positions of stars near the Sun should be shifted because of the curvature of space by the gravity of the Sun. Stars cannot be seen in daylight, but the eclipse provided the opportunity to observe this effect since the Moon blocks the light of the Sun. The aim was to measure the positions of stars near the Sun during the eclipse and to compare them with their positions in the night sky. On the morning of the eclipse there was a storm with heavy rain, but it cleared in the afternoon. Eddington was so busy taking photographs that he did not see the eclipse. But with the images he took he successfully showed that the prediction of Einstein was correct, and the results were reported in newspapers worldwide.

Until the beginning of the 20th century, our Sun was thought to be a burning ball of fire. But fires burn quickly and the astronomers calculated that our Sun could be no older than 20 million years.

Eddington began working on the structure of stars in 1916 when he calculated the importance of heat transfer in stars by radiation. Previously, it was thought that convection was the main process by which heat moved from the centre of stars to their edges. Then, in 1920, the same year as he published the results from the solar eclipse observations, he published a paper titled 'The internal constitution of stars'. In this work he discussed the age problem of the Sun, and suggested that a new previously unknown mechanism for generating energy must be at work. Eddington realised that the temperature and pressure at the centre of the Sun may be so high that hydrogen nuclei could literally fuse together, forming helium and fuelling the Sun via the conversion of a tiny amount of their mass into energy. With such an energy source he realised that a star like our Sun could shine for billions of years. He also stated that such fusion processes could create heavier elements inside stars[23].

Eddington had correctly identified the energy source of stars and a means to form many of the heavier elements, although the details took several decades to be worked out. He then derived relationships between the masses and luminosities of stars and published several influential papers on the life of stars, essentially marking the beginning of theoretical astrophysics. How stars evolve in time became one of the main research activities amongst astronomers and astrophysicists until the 1960s, by which time the basic workings of stars were finally understood.

When I am asked, what use is astronomy? My first response would be "does it need to have a use? Isn't discovering the cosmos enough with the inspiration it provides?" But then I will add, thanks to the work of Arthur Eddington who discovered the energy source of stars that we

---

[23] 'Arthur Stanley Eddington: pioneer of stellar structure theory', 2004, L. Mestel, Journal of Astronomical History and Heritage, vol. 7, pages 65-73.

will have a safe and virtually unlimited source of energy once nuclear fusion is harnessed.

The Sun is powered by the energy released when protons are forced close enough so that they form a larger atomic nucleus, releasing energy in the process. The mass of the resulting helium nucleus is not as much as the individual particles – it is about one percent less. That one percent of mass was lost when the strong force was overcome such that the protons could be squeezed together. This small amount of mass is equivalent to a large amount of pure energy - neutrinos and photons which provide the heat source emanating from the core of the star. These reactions between particles are very unlikely unless the conditions are extreme such as found at the centre of a star. Higher temperatures lead to faster-moving particles that collide more often and with more energy.

At the centre of our Sun, every second about 600 million tons of hydrogen are fused together to form helium releasing energy at a rate of 10^{26} Joules per second [24]. That is the equivalent of a trillion hydrogen bombs continuously detonating each second, providing the energy that makes the Sun shine. The central core of our star, about 20 percent of its radius, has a density about 150 times that of liquid water and a temperature of 15 million degrees. Because the core of our Sun

---

[24] Joules are our unit of energy, defined by the energy required to accelerate one kilogram to one metre per second per second through a distance of one metre. One joule is about the energy it takes to lift a small apple one metre above the ground, which will equal the kinetic energy released when the apple falls one metre to the ground (remember that energy is conserved). One watt, a measure of power or energy consumption, equals one joule per second.

is large, this is an energy of about 100 watts per cubic metre – equivalent to the energy used by a single bright light bulb.

You might wonder why the Sun does not just explode in a matter of seconds during a runaway chain reaction of fusion as the temperatures get hotter and hotter, creating a massive cosmic fusion bomb. This is because stars regulate themselves via a so-called feedback loop. If the temperature and pressure increase at their centre, the material gets pushed outwards and they start to expand. Expansion leads to cooling, rather like the cooling that occurs as the universe expands or as a gas expands. This then decreases the number of fusion reactions, and the star contracts again.

It takes a long time for the energy created at the centre of the Sun to reach its surface. The photons travel only a few millimetres before they collide with another particle which scatters the photon into a different direction. The photons make what is called a random walk, bouncing this way and that until sooner or later they end up at the surface of the star. Eventually they escape into space. This is a journey that takes up to one million years! The photons from the fusion process are high-energy gamma rays. But by the time they reach the surface of the Sun, they have transferred most of their energy to other particles. This keeps the interior of the Sun extremely hot. When they finally leave its surface, the photons emerge in the lower-energy optical wavelengths. The light you see from our Sun today originated at the centre of our star at a time before *homo sapiens* appeared, when our distant ancestors were beginning to use fire.

Eight minutes later the photons reach the surface of our Earth. The Earth's surface captures less than a billionth of the light coming from the Sun, but that is still a vast amount of energy. If we could capture all of the sunlight that strikes the

Earth's surface in a single hour, we could provide the global power requirements of the Earth for one year. In fact, the total energy output of the Sun could support 20 trillion planets like our Earth for 15 billion years. That is an impressive amount of energy, but we would need a construction called a Dyson sphere to capture it all[25].

In 1926 Eddington published an influential book 'On the internal constitution of stars'. This contained many new insights into the structure of stars but I want to mention just one of those. By treating a star as a sphere of gas, he derived simple theoretical relationships between the luminosity, mass, radius and temperature of a star. He then applied this theory to a strange new class of stars that were puzzling astronomers – white dwarf stars. If you recall Friedrich Bessel observed the motion of the star Sirius which suggested that it was orbited by another object. By the beginning of the 20th century, telescopes had resolved the extremely faint companion object and measured its spectrum of light. When Eddington calculated the density of this object, he found that it was 53,000 times the density of water. This was extremely puzzling since the average density of the Sun is not much higher than that of water - even gold has a density of just 20 times that of water. The puzzle was soon solved thanks to the new theories of quantum mechanics which showed that matter could be squeezed to much higher densities. Many scientists contributed to the understanding of these strange white dwarf stars over the following decade. It turned out that the white dwarf stars were the remnants of stars like our Sun. At the end of its life, when the Sun stops fusion, gravity takes over and the remaining matter collapses into an object the size of our

---

[25] We will learn more about Dyson spheres in chapter 40 on Freeman Dyson.

Earth. Initially hot from this collapse, the newly formed white dwarf will slowly cool down and fade away.

Eddington died after an operation in a nursing home in 1944, several months after the start of World War II. He would have been appalled at the detonation of the first atomic bomb in 1945, and even more so by the first hydrogen fusion bomb in 1952. After all, this most powerful and horrific weapon followed from the research of pacifists Einstein and Eddington. The most destructive weapon ever made by our species was the Soviet Tsar Bomba which was exploded in 1961 on a remote island off the northern coast of Siberia. It was equivalent to detonating 50 million tonnes of the explosive TNT (which would fill a cube 300 metres on each side) and was over a thousand times the combined power of the two nuclear fission bombs used to destroy Hiroshima and Nagasaki in the Second World War. If you could contain the energy from this single bomb and convert it into electricity, it would have been enough to power a small country for a year. Nuclear fusion is a very clean energy source since it does not use or release radioactive elements. Harnessing this energy is difficult since you have to recreate the conditions at the centre of a star. However, in 2021, China's EAST fusion experiment heated a plasma to 120 million degrees for over a minute. It is thought that turning these experiments into working energy generating power stations will take a few more decades.

Once we take this new energy source into consideration in our calculations, we realise that the Sun has an enormous reservoir of fuel. There is sufficient hydrogen in the core of our Sun for it to shine for ten billion years. Thanks to Eddington, the debate between astronomers and geologists and biologists was finally put to rest. Moreover, we have calculated that our

Sun has been shining for 4.6 billion years[26]. This is slightly older than the age of the oldest rocks on Earth and the Moon, but similar to the oldest meteorites. This is not a coincidence since we believe that the Sun began shining at a time when the building blocks of planets were forming, and the planets formed soon after.

Eddington was nominated six times for a Nobel Prize between 1932 and 1944. However, even by 1944 no astronomer, astrophysicist or even cosmologist, had been awarded the prize. It was only in 1967 that the Nobel Prize in physics was awarded for work in astrophysics - to the physicist Hans Bethe for expanding on Eddington's ideas! His award sentence states: *"for his contributions to the theory of nuclear reactions, especially his discoveries concerning the energy production in stars."*

Despite numerous nominations and brilliant discoveries by scientists in these domains, why were these entire research disciplines not considered for so many years?

Bethe recounts the following story in his text 'My life in astrophysics'[27]. During lunch before the prize ceremony, he recalls a conversation with a 'Mr Rydberg' (presumably Erik Rudberg, chairman of the physics Nobel Committee): *"Perhaps the most interesting to me was the reason that I was the first to be honored for work in astrophysics. Nobel's wife, he told me, had run off with one of the leading mathematicians and astronomers of the time. So the Prize bequest had specified that the work honored had to*

---

[26] The Sun is the only star for which we can directly determine a precise age, thanks to additional information we can obtain regarding its composition and internal structure from helioseismology. But we can determine the ages of clusters of stars.

[27] Annual Review of Astronomy and Astrophysics, 2003, 41, page 7.

have practical application and that neither pure mathematics nor astronomy could be considered." This anecdotal story has often been repeated but there is no additional evidence to support this – and Alfred Nobel was never married, although he did have several lovers.

Helge Kragh, a renowned historian of astronomy, researched the Nobel Prize archive of nominations and committee deliberations to try to uncover why astronomy and astrophysics were neglected during the first six decades of the Nobel Prize[28].

The physics Nobel Prize committee normally consists of five distinguished scientists. Kragh discovered from the Nobel archives that in 1923, the majority of this committee, Carl Oseen, Manne Siegbahn and Svante Arrhenius, argued that astronomy was too far from physics to be considered for an award. Furthermore, they considered astrophysics, and thus cosmology, too similar to astronomy. For the following four decades this precedent left numerous incredible scientists and their discoveries unrewarded, including Eddington.

---

[28] The Nobel Prize System and the Astronomical Sciences, Helge Kragh, Journal for the History of Astronomy, 2017, Vol. 48, page 257-280. The archives were made available 50 years after the fact, although in 2017 they seem to have lengthened this timescale. As of 2022 the nomination archives are only available up until the year 1966.

16. Cecilia Payne (1900-1979)

"for discovering what stars are made of"

In 1921 Arthur Eddington was giving a talk at Cambridge University describing his expedition to observe the deflection of star light to test Einstein's theory of general relativity. Attending the lecture was a young student studying botany, Cecilia Payne. This inspired her to follow a career in astronomy, and within a few years she would discover what stars are made of.

Cecilia Payne was born in England in 1900. At St Paul's Girls school in London, she was urged by her teacher, the composer Gustav Holst, to pursue a career in music but she wished to study science. She won a scholarship from Newnham College, one of two colleges where women were allowed to live and attend lectures at Cambridge University. She began studying botany until she heard a lecture on general relativity. The college had received four tickets to Eddington's lecture and Payne was fortunate to receive one of those tickets after another participant dropped out. After she heard Eddington's speech, she immediately switched to studying mathematics, physics and astronomy. In her autobiography, 'The Dyers Hand', she wrote *"The result was a complete transformation of my world picture. [...] My world had been so shaken that I experienced something very like a nervous breakdown."*

Despite completing her studies women were not allowed to obtain degrees from Cambridge until 1948. They could attend lectures but had to suffer in the process. The men had tutors assigned to them but not the women as there was no woman qualified to tutor in physics. Payne was the only female student studying physics at the time. At Cambridge the women were not allowed to sit in the same row of seats as the men and she had to sit at the front of the lectures alone[29].

Despite the support and encouragement of Eddington, Payne realised that she could not follow her dreams of becoming an astronomer in England – there was simply no career path for women in such research. Fortunately, Eddington had recognised her talents and recommended Payne to Harlow Shapley, the director of Harvard University Observatory. Shapley agreed to provide her with some funds for a year or two and she left for America in 1923. Within two years she had completed her thesis which radically changed our understanding of stars and their composition. Shapley, like many PhD advisors today, liked to say that students only deserve a PhD if they suffer in the process. Indeed, Payne wrote in her autobiography *"There followed months, almost a year as I remember, of utter bewilderment. Often I was in a state of exhaustion and despair, working all day and late into the night"*.

That stars contain a range of elements was first determined using the method of spectroscopy. It was Isaac Newton who first passed the light of the Sun through a prism and noticed the spread of colours. This observation put to rest a debate about the origin of the colours in a rainbow that began, probably when humans first looked at a rainbow, but at least as early as Aristotle. By the beginning of the 19th century the German astronomer Joseph Fraunhofer found that the Sun's spectrum of light contained numerous dark bands. He also noticed similar bands in the spectrum of light from distant stars when their light was focused with a telescope onto a prism. He had constructed the first 'spectrograph' for these experiments which used a diffraction grating, a sheet with

---

[29] 'Cecilia Payne-Gaposchkin', interviewed by Owen Gingerich 1968, American Institute of Physics Oral History Interviews

hundreds of narrow slits, in front of the prism to disperse the wavelengths of the light. About 45 years later, the German scientists Gustav Kirchhoff and Robert Bunsen noticed that the position of some of Fraunhofer's dark lines coincided with the characteristic emission lines of light that would arise when different elements were heated. They correctly inferred that the dark lines in the solar spectrum are caused by the absorption of light by chemical elements in the solar atmosphere. The element helium was first discovered as an unknown dark band in the Sun's spectrum in 1868.

Some of the most prominent absorption lines in the spectrum of the Sun are due to elements such as iron, silicon and oxygen. At the beginning of the 20th century, it was naturally assumed that these elements were the most abundant in the Sun and that stars were made of the same mix of elements as the Earth. The well-known Princeton astronomer, Henry Russell, had stated that if you heated the Earth's crust to the temperature of the Sun, its spectrum of light would look the same. At the same time, observations of the spectra of numerous stars were being carried out at Harvard University Observatory. Another of Edward Pickering's human 'computers' was Annie Jump Cannon. She was hired to classify stars according to their spectra. She created a classification scheme for stars that astronomers still use today, which groups stars in terms of their colours and spectra. Astronomers naturally assumed that the differences in the star's absorption spectra were due to differences in their composition, an assumption that Payne showed to be incorrect.

The emerging research into quantum mechanics showed how atoms could only absorb light at specific energies that correspond to the difference in the possible energy levels that their electrons could attain. In 1920 the Indian astrophysicist Meghnad Saha had calculated that in the outer regions of stars, elements can have one or more electrons stripped away and the remaining electrons could be in a variety of energy levels depending on the conditions on the surface of the star. These are called ionisation states of an atom. In India, Saha did not have access to the data to take his studies further. Payne used the ideas of Saha and applied them to dozens of detailed spectra of stars that were available at Harvard Observatory. She calculated that the differences in the spectrum of light from different stars were not due to a variation in the composition of stars, but were due to different ionisation states of the elements that arise because of a variation of the surface temperatures of the stars. She went on to show that the data were perfectly consistent if all stars had a similar chemical composition.

This was a profound discovery that changed our understanding of the structure of stars. The uniformity of stars and their compositions was the central focus of her PhD thesis – a 200 page book that was later described by Otto Struve, the director of Yerkes Observatory, as *"the most brilliant PhD thesis ever written in astronomy"*[30].

There was more. Another chapter of her thesis was titled 'The relative abundances of the elements'. Payne applied her theoretical models of the ionisation states of the elements to estimate their abundances. She realised that in the atmospheres of stars, only about one in 200 million of the hydrogen atoms should be in a particular excited state that was giving rise to a particular hydrogen absorption line. The hydrogen you see in this ionisation state was just the tip of the

---

[30] 'Cecilia Payne-Gaposchkin: astronomer extraordinaire', P. A. Wayman, 2002, Astronomy & Geophysics, vol. 43, pages 27-29.

iceberg! Using a similar argument for helium she discovered that hydrogen and helium were by far the most abundant elements in stars, and therefore in the universe as a whole. This was an incredible discovery. When astrophysicists such as Eddington were working out the details of how stars evolve in time he assumed, as did all astronomers at the time, that stars were made mainly of iron and silicon like our Earth, and that only a small fraction of the star was made of hydrogen and helium.

When she submitted her thesis, Henry Russell was her external thesis reviewer and wrote *"It is the best doctoral thesis I have ever read."* However, before publishing her thesis Payne had sent her results on the abundances of the elements to Russell who wrote Payne a letter stating *"I am convinced that there is something wrong seriously wrong with the present theory. It is clearly impossible that hydrogen should be a million times more abundant..."* This might sound harsh, but this is often how science works – especially with radical results, your work will be seriously questioned. Moreover, Russell was a famous astronomer and Payne was a young researcher and a woman in a research world completely dominated by men. Discouraged, she wrote in her thesis that her results must be in error. But she wisely left her findings in the thesis rather than removing them.

In 1929, Russell himself applied similar techniques to study the composition of the Sun and found the same result – that hydrogen and helium were indeed the most abundant elements. Despite giving Payne credit for the discovery in his own publication, for the next few decades astronomers associated Russell with this discovery. I looked through dozens of scientific papers on the topic from 1930 until 1980 and rarely found a mention of Payne's work. As often is the case, the most famous scientist at the time receives the credit for a discovery, even though the evidence is plainly there in the scientific publication record as someone else's work.

No woman had previously received a PhD in physics at Harvard – at the time there was no graduate program in astronomy. But the head of the physics department signed the approval for Payne's thesis from Harvard's Radcliffe College, which essentially marked the beginning of the Astronomy Department at Harvard. She would continue her research as an assistant to Shapley, and published studies on the exotic elements found in the most luminous stars of our galaxy.

In 1933 Payne travelled to Europe and went as far as Leningrad to visit astronomers working there. She met the astronomer and Russian exile, Sergei Gaposchkin, whom she would marry the following year and take on the name Payne-Gaposchkin. Together they carried out a systematic study of variable stars in the nearby Magellanic Cloud galaxies, making more than a million visual estimates of their properties. It was only in 1938 that Payne received the title 'astronomer' at the Harvard Observatory. In 1956 Payne became the first female to advance to full professor within the Harvard Faculty of Arts and Sciences and the first female to direct the Astronomy Department. She would go on to write several hundred scientific articles and several technical and popular science books.

Thanks to Cecilia Payne we now know that by mass, the composition of the Sun is similar to most stars and is 74.9% hydrogen 23.8% helium and only about 1.3% of its mass coming from all the other elements. This is similar to the composition of the so called 'interstellar medium', the gaseous component of galaxies which contains the recycled elements of stars and within which new stars are forming today. In the next chapter we will hear the story of how George Gamow explained why this characteristic ratio of hydrogen and helium arises in the early universe.

In the early 20th century one out of every three American astronomers was a woman, but their names are conspicuously absent from the textbooks and histories of astronomy. Most of them had careers lasting less than five years. Like Henrietta Leavitt and Annie Jump Cannon, they were primarily hired to do mundane tasks such as analysing photographic plates. Payne was one of the first successful astronomers to have a career whilst being married and having children. At the time many institutions did not even allow women to continue working if they married. Use of the great telescopes was restricted to male astronomers until the 1960s. For an informative view of this period, I recommend the short article by George Greenstein 'The ladies of Observatory Hill: Annie Jump Cannon and Cecilia Payne-Gaposchkin'[31].

Payne was a chain smoker and died of lung cancer aged 79. The obituaries in the New York Times and Washington Post stated that she was *"a pioneering astrophysicist and probably the most eminent woman astronomer of all time"*, but they did not even mention what she discovered – the substance of stars. Even the obituaries in the science journals did not give her full credit for her research. This was first remedied by the historian of science, Owen Gingerich, who credited her properly in his 1982 obituary in the Quarterly Journal of the Royal Astronomical Society.

Cecilia Payne had overcome numerous obstacles to follow her passion and fully deserved the award of a Nobel Prize for

---

[31] "The ladies of Observatory Hill: Annie Jump Cannon and Cecilia Payne-Gaposchkin", 1993, in The American Scholar, Vol. 62, no. 3. pp 437-446.

which she was never even nominated. She wrote in her autobiography that *"...being a woman has been a great disadvantage. It is a tale of low salary, lack of status, slow advancement. [...] It has been a case of survival, not of the fittest, but of the most doggedly persistent."* She ends with some advice for young researchers and particularly young women: *"Do not undertake a scientific career in quest of fame or money. There are easier and better ways to reach them. Undertake it only if nothing else will satisfy you; for nothing else is probably what you will receive. Your reward will be the widening of the horizon as you climb. And if you achieve that reward you will ask no other."*

17. George Gamow (1904-1968)

"for the theory of big bang nucleosynthesis"

Although the ideas of Lemaître were known at the time, they were not widely accepted as proof that the universe had a hot and dense beginning. Interest in the idea of the big bang was reignited in the 1940s by George Gamow. He was the first to apply the emerging knowledge of nuclear physics to astrophysics and cosmology which revolutionised our understanding of the workings of stars and provided evidence for the hot big bang origin of our universe.

Gamow was born in Odessa, which is now in the Ukraine but was then part of the Russian Empire. At the age of six he climbed on to the roof of his house to get a clear view of Halley's comet. His parents were school teachers and he was mostly self-taught and never formally obtained a PhD. He moved to Leningrad in 1923 to become a student of Alexander Friedmann, who had first demonstrated that Einstein's equations have solutions that could be applied to a universe that evolves in time. Unfortunately, Friedmann died in 1925. Left without a PhD supervisor, Gamow became interested in the new theory of quantum mechanics after reading a publication of Erwin Schrödinger.

In 1928 Gamow was attending a summer school in Göttingen where he was studying the work of Ernest Rutherford on the subject of radioactive decay. The problem at the time was understanding how particles could escape from the nucleus. With the quantum mechanical description of particles just two years old, Gamow had taught himself the theory and applied the new probabilistic wave mechanics to the decay of particles. He showed that there is a chance that a particle exists in a region where the old classical view of particles would not allow it to be. Within a day he had formulated the first detailed mathematical explanation of radioactive decay of elements by 'quantum tunnelling'.

At the end of the summer school he only had ten dollars left, but wanted to travel to Copenhagen to meet the famous physicist Niels Bohr and tell him about his results. Upon hearing Gamow's new ideas and learning that he only had enough money to stay one night, Bohr offered Gamow a one year fellowship to stay in Copenhagen which he immediately accepted. After Copenhagen he was offered another fellowship to work with Rutherford in Cambridge before he finally returned to Leningrad in 1931.

During the 1920s Gamow had been allowed to travel freely and he visited many of Europe's top research institutions for work. But in the 1930s travel became restricted and Soviet oppression grew. Gamow attempted to escape the Soviet Union by canoeing 170 miles across the Black Sea to Turkey[32]. He and his wife spent three days in a storm and the boat was thrown ashore 100 kilometres from where they started. Luckily the state officials did not hear of this plan and later approved that he could attend the well-known Solvay physics conference in Brussels. This gave him the opportunity to defect to the United States. In the late 1930s he began applying his nuclear physics knowledge to the working of stars and constructed theoretical models for how stars evolve in time.

Gamow was a scientist with many original ideas and moved from topic to topic. He was the first to consider the conditions during the early stages of the big bang – a hot dense primordial sea of interacting particles within Lemaitre's primeval atom. Gamow realised that the conditions in the early universe were even more extreme than what was

---

[32] 'George Gamow', interviewed by Charles Weiner 1968, American Institute of Physics, oral history interviews

occurring at the centres of stars. He started with the assumption that the universe was filled with neutrons and photons. As the universe expanded and cooled, a series of interactions would take place, first forming protons, and then some of the heavier elements, as protons and neutrons combined together. He published his ideas in 1946 together with his PhD student Ralph Alpher and was able to predict the origin and abundance of hydrogen and helium, the most common elements in our universe.

Gamow and Alpher had showed that hydrogen and helium could be produced in the early universe, but few other elements could arise. This was a problem at the time since the origin of the heavier elements was left unsolved. However, the ratio of hydrogen and helium predicted from such calculations is exactly the ratio that is observed in our universe, providing very strong evidence for the big bang model and the fact that the universe was once dense and hot enough to create light atomic nuclei. For every helium atom there should be around four hydrogen atoms. This prediction only depends on the slight difference in masses of the neutron and proton. It does not depend on the details of the big bang, only that the universe was once hot and dense. There is no alternative model for our universe that can give rise to the observed abundance of hydrogen and helium. However, at the time of Gamow most astronomers were rather hostile to these ideas from a theoretical nuclear physicist and the idea of an expanding universe from a hot big bang was still not generally accepted.

Gamow was a notorious prankster. In his published paper with Alpher on the synthesis of elements during the big bang, 'The origin of the chemical elements', Gamow added the name of his friend and colleague Hans Bethe to the author list. Bethe had not seen the paper and Gamow added his name as a joke for the sole reason that the authors' surnames would sound like the first three letters of the Greek alphabet 'alpha, beta, gamma'.

Gamow once submitted a joke paper to the prestigious journal Nature, using the physics of the rotation of the Earth to explain his observation that cows in the Northern hemisphere chewed clockwise and those in the Southern hemisphere chewed anti-clockwise. In 1940 he published a paper with the Brazilian physicist Mario Schoenberg about the role of neutrinos in massive stars. They named the neutrino reaction 'the Urca process' after a well-known Rio de Janeiro casino where the two scientists first met.

Despite his humorous approach to life, his insights were often very accurate. Later in his career Gamow became interested in biology. In a lengthy interview about his life's work a few months before his death in 1968 aged 64, he was asked about his most important work. Gamow replied that probably it was his work on the genetic code of life. In 1954, he was the first to correctly propose that sets of three nucleobases could encode the 20 amino acids used by all life to construct proteins.

Thanks to the pioneering work of Gamow, we now understand the history of our universe all the way back to the time of a fraction of a second after the big bang. He published six technical books, 19 popular science books and over 100 scientific papers. Gamow was nominated twice for a Nobel Prize, in 1943 and 1946, but not only did he not receive this recognition, he never received a single award during his career – apart from the UNESCO Kalinga prize in 1956 for his work in popularising science. At a conference devoted to his research thirty years after his death, his colleague Robert

Herman confirms the widespread belief that it was his outlook to life and his humour that probably led to his lack of recognition, writing *"It is rare to encounter a man who was so incredibly creative, imaginative and who enjoyed life so much. I dare to say that too many of his scientific contemporaries took umbrage at his enjoyment of life and of physics. I think that this was a tragedy, and it would have been nice to have seen him receive much more of the recognition that he so richly deserved."*[33]

---

[33] George Gamow Symposium, 1997, ASP Conference Series, Vol. 129.

18. Ralph Alpher (1921-2007) & Robert Herman (1914-1997)

"for predicting the measurable afterglow of the big bang – the cosmic microwave background"

The third pillar of evidence for the big bang was the discovery of a sea of photons that originated during the early universe and still bathes us in light today – the cosmic microwave background. The existence of this primordial light was predicted by Gamow's student, Ralph Alpher together with his colleague Robert Herman. It was quite a feat to predict the existence of phenomena that would not be observed until several decades later. And its discovery provides the strongest evidence that our universe began in a hot and dense fireball.

Ralph Alpher obtained his PhD in 1948 at Johns Hopkins University Applied Physics Laboratory, his dissertation being his important contributions to the infamous Alpher, Gamow & Bethe paper. Alpher was not happy with Gamow's joke about including Bethe on the paper – he rightfully thought that his own contributions would be ignored since people would only see the names of the famous Gamow and Bethe. Straight after his dissertation he began to calculate the evolution over time of the hot radiation that was present in the early universe, together with Robert Herman whose office was just a few doors away.

Robert Herman obtained his doctoral degree in physics from Princeton University in 1940, and after the Second World War he joined the Applied Physics Laboratory. Alpher and Herman spent a great deal of time discussing science together. Herman carefully went through Alpher's PhD dissertation with him and they began discussing the many consequences. One of those was the question of what happened to the photons that were present during this era. To work out the details required a deep understanding of mathematics and physics – the pair worked together in front of the blackboard, correcting each other and advancing the calculations until they came up with the answer. The afterglow of the big bang should still be present today, permeating all space and reflecting the average temperature of today's universe that is just a few degrees above absolute zero.

During the early universe, particles of matter and antimatter collided and annihilated each other, releasing energy in the form of photons. After all the antimatter had annihilated with matter, an excess of matter was left together with a vast number of high-energy photons – the afterglow of the beginning of our universe. After 380,000 years the temperature of the universe had cooled to a modest three thousand Kelvin, less than the surface temperature of the Sun today. At this time the universe literally glowed in light that would be visible to our eyes with a colour temperature similar to a 'warm white' light bulb. The energies of electrons had decreased sufficiently to allow them to bind to the atomic nuclei and neutral atoms of hydrogen, helium and lithium formed. Prior to this time the photons were colliding with and bumping off the nuclei and electrons. Quite suddenly, the photons could move freely past the neutral atoms. From this time onwards, called decoupling, the photons no longer had anything with which to interact, and they continued their timeless journeys unimpeded, moving in all directions. These are the cosmic microwave background photons.

Since the big bang photons were emitted, the universe has grown in size a thousand times larger. This stretching of space also stretches the photon wavelengths, so rather than appearing a warm white, today they would be microwave photons with far less energy. "Where did that energy go?" you may ask. It's a good question. The theory of general relativity does not say anything about whether or not energy must be conserved. However, I like to interpret this as the energy of the photons being transformed into the gravitational potential energy of the entire universe – after all it takes energy to expand space and this energy must come from the contents which cause this expansion.

Most scientists thought that this radiation would be too weak to detect, and its prediction did not seem to catch the attention of experimentalists at the time, despite numerous colloquia on the topic given by Alpher and Herman. One of the problems, recalls Alpher and Herman in a 1983 interview[34], was that they estimated that the energy intensity of the background radiation today was comparable to the energy intensity in all the background star-light. If they had mentioned in their paper that the radiation would have wavelengths stretched into the microwave frequencies, it might have been picked up on by the radio astronomers. The cosmic microwave background was discovered by accident by at least four independent research groups in the 1950s and early 1960s. It is a rather strange story how Arno Penzias and Robert Wilson received all the credit for its discovery in 1964[35].

Penzias and Wilson, working at Bell Research Laboratories, were constructing a telescope to observe the universe in the radio wavelengths. They were frustrated by a constant background noise from their antenna which seemed to come from every direction in which they aimed the telescope. At first, they thought the noise was due to the pigeons that nested inside the horn-shaped antenna. Cosmologist Jim Peebles pointed out to them that they might have observed the signals of relic photons from the big bang. Peebles had himself just

---

[34] American Institute of Physics, Oral History Interviews, Ralph Alpher and Robert Herman – Session II, 1983.

[35] 'Early photons from the early universe', V. Trimble 2006, New Astronomy Reviews, 50, 844-849.

calculated the existence of the microwave background photons. Initially he was unaware of the work of Alpher and Herman. Upon submitting his study for publication, Herman received it to referee and pointed out that there was nothing new in the paper since the results were all in their 1948 study. The paper was duly rejected. Peebles submitted his work to a different journal where it was deliberately published next to the discovery paper of Penzias and Wilson in 1965. It still did not contain a reference to the original work of Alpher and Herman.

Penzias and Wilson received the Nobel Prize in physics in 1978 for their discovery. At the time many complained that Peebles and his co-author Robert Dicke should have also shared this prize. In reality, it was Alpher and Herman who should receive recognition for the theoretical prediction. In their 1983 interview, Alpher and Herman reveal their distress at their work being ignored despite many scientists being aware of it. In a bizarre twist to the story, it turned out that Penzias and Wilson were not the first to detect this radiation by accident!

In 1940 the Canadian astronomer Andrew McKellar observed the absorption of light by interstellar cyanogen gas. From the way in which light was blocked at two different frequencies he calculated the temperature of interstellar space as 2.3 Kelvin, very close to today's measured temperature of the cosmic microwave background of 2.7 Kelvin. The cyanogen gas is kept warm by the microwave background photons. Soviet radio astronomer Tigran Shmaonov reported in his 1957 PhD dissertation that he mapped the sky in the microwave frequencies using a microwave antenna – similar to that used by Penzias and Wilson. He reported a radiation background of 4 Kelvin in all directions. Also in 1957, the

French radio astronomer Emile Le Roux surveyed the sky using a World War II German radar dish and found the temperature of the sky to be 3 Kelvin. He suggested an extragalactic origin in his work. In 1961 Edward Ohm published his results from radio maps of the sky and clearly detected an excess temperature of 3 Kelvin. Ironically, Ohm had gathered their data using the same antenna Arno Penzias and Robert Wilson used at Bell Laboratories. None of these authors claimed to have detected the big bang photons, but they had, just like Penzias and Wilson, discovered them by accident.

In yet another twist to this story, Soviet scientists Andrei Doroshkevich and Igor Novikov published a paper in 1964 showing that the relic radiation should be detectable in the microwave frequencies. They also mentioned that the Bell Laboratories radio antenna was ideally suited to its detection, and they looked at the 1961 observations of Ohm, but misread the results and concluded that this contradicted the big bang theory! If they had not misread a particular notation used by Ohm, probably Ohm would have been awarded the Nobel Prize rather than Penzias and Wilson.

Several subsequent experiments, confirmed that the universe was filled with low energy photons with a temperature of 2.725 Kelvin. But it was not until 1990 that everyone became convinced that these photons must have originated from the big bang. The cosmic microwave background photons were predicted to have a characteristic energy distribution that is detectable in the radio and microwave bands from 0.03 to 300 centimetres, and that is exactly what was observed by the NASA's Cosmic Background Explorer satellite in 1990. The COBE satellite also observed tiny temperature variations in the cosmic photons which reflected small variations in the distribution of matter in the early universe. Another two Nobel Prizes were given for this discovery, to George Smoot and John Mather in 2006.

In 1990 the evidence was so strong in support of the big bang model of the universe that it immediately put paid to all competing theories. The most compelling of these was Fred Hoyle's steady-state model, but it too could not explain the observed photon background. Despite many subsequent attempts at alternative cosmological theories, none come close to matching these data and most cannot even explain the presence of these photons in the first place.

I am often asked where these photons come from. During the early universe when matter was annihilating with anti-matter, photons were being produced at every point in space. These photons began travelling from every place in the universe. The photons arrive at our detectors after travelling for nearly 13.8 billion years, reaching us from the edge of our observable universe. The glow of light is the same in all directions. This confirms the Copernican Principle, that our universe is the same in all directions.

This leads to the follow up question, "why haven't most of these photons simply travelled past us already?" In fact, they have. The cosmic microwave background photons we see have been travelling towards us for the age of the universe. Those photons that were generated at closer distances than the horizon, the furthest distance we can see, have indeed passed us. We essentially see the light from a thin shell that lies in all directions at the edge of our visible universe 46 billion light years away. The light reaches us because most of space is empty and there is little matter to block its path. As time proceeds, we see photons that originated from slightly further distances. From the regular nature of this light in all directions, we also can infer that the universe is quite a bit larger than what we see. Observers in different parts of the universe will see photons originating from the edge of different large horizon-spheres, of the same size but centred on their location.

While at John Hopkins Research Laboratories, Alpher and Herman published 11 scientific papers together, mostly about the early universe, but not all were. In a 1951 joint study they used the abundance of lead in Earth's crust to place an upper limit to the age of the Earth of 5.3 billion years. In another paper in 1953 they predicted the existence of a cosmic neutrino background that resulted from the annihilation of particles just one second after the big bang. This would be extremely hard to directly detect as it would be at an even lower temperature than the photons today, however experiments are being constructed to try to detect its presence. If successful, more Nobel Prizes will probably be awarded.

In 1955 Alpher went to work for General Electric Research Laboratory and in 1956 Herman left to work at General Motors Research Laboratory. They were allowed to spend five percent of their time on 'other research' and the two often talked and collaborated on ideas related to the early universe. They continued to write a few papers together, their last was in 1998 on the density of the universe. The popular science book they started writing together titled 'Genesis of the Big Bang' was published in 2001, four years after the death of Alpher. It contains many personal reminiscences which must have been painful for the authors.

When it was announced that Penzias and Wilson had discovered the cosmic radiation, Ralph Alpher was at first ecstatic, but then he was shocked to find not a single reference to his work. He tried to set the record straight, writing to many scientists. When Penzias and Wilson were awarded a Nobel Prize in 1978, Penzias approached Alpher to get his version of the story. Alpher recounts *"I spent a day and a half with him in which I gave him a crash course on cosmology, and he didn't know a damned thing."* A month later Alpher suffered a heart attack which he partly blamed on the stress of that meeting. However, Penzias gave a clear and accurate account of the history in his Nobel Prize acceptance speech crediting Alpher and Herman for their discovery. But still, most cosmologists continued to give credit to the wrong people. Even Stephen Hawking received a letter from Alpher, pointing out that in his famous book 'A brief history of time', his work was not mentioned and credit was incorrectly given to Gamow and Peebles.

In a 1999 interview for Discover magazine, titled 'The last big bang man left standing', Alpher expresses his lifelong disappointment about not being recognised for his research. *"The important thing is to get the credit in the literature,"* he says. *"There are two reasons you do science. One is an altruistic feeling that maybe you can contribute to mankind's store of knowledge about the world. The other and more personal thing is you want the approbation of your peers. Pure and simple."*[36]

Robert Herman seems to have been less bothered by the lack of recognition. He was described by his colleagues as a gentle and modest man as well as a creative genius with contributions to art, science and engineering. Whilst at General Motors he invented the research area of traffic science, applying physics to traffic flow. His work on cities and

---

[36] I think these sentiments can be appreciated by all my colleagues, after all, whenever we first release a publication, the first emails we receive are inevitably from scientists who inform us that we have neglected to cite their work!

networks helped inspire the popular video game series 'Sim City'. He was passionate about collecting and playing antique cellos and applied his knowledge of physics to the mechanics of the cello bow. With his joyous outlook on life he was much loved by his students and colleagues. For Herman, it was the pursuit of truth that drove him, not accolades and he stated *"You don't give recognition to the person, you give it to the work."*

19. Clair Patterson (1922-1995)

"for the measuring the age of the solar system"

Clair Patterson was born in 1922 in rural Iowa, America. When he was a child, he had asked his mother *"why are water drops round?"* His mother, a school teacher, bought him a chemistry set so he could set out to answer such questions. He had soon constructed his own laboratory in the basement of his parents' house. He would later become professor of chemistry, discover the age of the solar system and save countless lives in the process!

The first scientific estimates of the age of the Earth were made by assuming that it was very hot when it formed and that it slowly cooled down by radiating heat until it reached the temperature it has today. William Thomson (Lord Kelvin), originator of the fundamental temperature scale, used this technique to estimate its age and came up with a timescale that was just tens of millions of years. This led to intense debates with the geologists and biologists who were studying processes on our planet which seemed to require hundreds of millions of years to take place.

Although the physicists made a correct calculation based on the known physics of the time, the estimates were wrong since there were additional unknown heat sources at work that had yet to be discovered. It was also not known that convection could bring heat from Earth's core to the surface. Scientists were also not aware of the heat that radioactivity could supply until the work of Henri Becquerel and Ernest Rutherford. The Earth contains a small fraction of unstable elements which decay on a several-billion-year timescale, releasing energetic particles that are absorbed by the surrounding rocks and thus increasing their temperature. Volcanoes and lava flows, plate tectonics and thermal hot springs are all powered mainly by radioactivity within the Earth's upper mantle. By not including this heat source within their models of a cooling Earth, scientists had obtained the incorrect answer for its age.

The discovery of radioactivity not only gave a way to keep the Earth hot, it led to a new and precise way of dating things like fossils or rocks. We can analyse their chemical composition and use 'radiometric dating' to determine their age. In a nutshell, the relative numbers of certain atoms change in a well-defined way over time and by counting the number of mutated atoms we can date the object.

Consider uranium-238, so called because it has 146 neutrons and 92 protons. It has a half-life of about 4.5 billion years, meaning that there is a 50:50 chance of a single atom changing to lead within this timescale. If a rock containing this form of uranium formed 4.5 billion years ago, today half of the atoms will have turned to lead and half will remain unchanged. Four-and-a-half billion years into the future, one-half of these remaining uranium atoms will have also mutated. In other words, three quarters of the original uranium atoms will have become lead. By counting the ratio of lead atoms to uranium atoms, we can determine the age of the rock. The mineral zircon is typically used to date the oldest rocks since as it forms it incorporates uranium into its crystalline structure but not lead. Any lead atoms found in zircon today must have originated from the decay of the uranium atoms that were trapped inside its structure when it first crystallised.

It was the American chemist Bertram Boltwood who discovered the technique of age dating using the decay of uranium to lead. In 1907 he published results with rocks as old as 2.2 billion years. Since the oldest rocks on Earth may have been destroyed over geological time, it was not clear just how old our Earth, let alone the solar system, could be. That is, until 1953 when Clair Patterson first measured its age.

Patterson studied chemistry at Grinnell college and obtained his degree in 1943. He then began graduate studies at University of Iowa. As the Second World War was well underway, he wanted to join the army but was persuaded to work on the Manhattan project to help construct the atomic bomb. He became part of the research group that had to separate the uranium-238 isotope from the uranium-235 needed for the bomb. This was done with a mass spectrometer that could measure the masses of individual atoms and then separate out the two isotopes. When the war ended, he returned to the University of Chicago to carry out his PhD, where his experience with mass spectrometers would be applied to dating ancient rocks. This required measuring the abundances of lead that were a thousand times lower than anyone had measured before.

Patterson applied radiometric dating techniques to meteorites to date the age of the solar system. He made the correct assumption that meteorites were the left-over debris from the beginning of the solar system and the formation of the planets. He also correctly assumed that if they had been orbiting in the vacuum of space their chemical composition would be unchanged over time. Patterson set out to measure the age of zircon crystals within meteorites. However, he found that these would quickly become contaminated with atmospheric lead when exposed to air. He also noted that the lead content of a single human hair could contaminate his experiments. In order to avoid contamination, he created the world's first sterile laboratory in 1952. He installed air filters and wore special clean suits and did not allow anyone into his laboratory. After many years of careful research, he measured the age of the solar system to be 4.55 billion years with an error of just 70 million years. Seventy years later this value has not changed – further research has only narrowed the margin of error on this age.

Patterson was curious as to the source of the atmospheric lead that was contaminating his meteoritic samples. He had a hunch that the cause was the lead used in automobile petrol to increase engine performance. Lead does not decay, and it accumulates in the air, water and in our bodies. And it stays there forever just like it stays in the meteorites he analysed. To test his theory, he obtained funding from the American Petroleum Institute to analyse the lead content in ocean sediments. The oil industry funded his research because they thought it would help them identify oil deposits. Patterson found that the ancient sediments had a fraction of a percent of the lead levels that were in the rivers of water feeding the oceans. He also found that the upper ocean water had a much higher lead fraction than the deep ocean. This was consistent with his hypothesis of recent atmospheric lead contamination, since it takes hundreds of years for the upper ocean waters to mix with the lower ocean. He estimated that the amount of contamination was consistent with the amount of lead emitted from car exhaust fumes. When he published his results in 1963, the Petroleum Institute withdrew his funding and even tried to block other sources by which he could fund his research[37].

Patterson managed to obtain government funding so that he could travel to Greenland and Antarctica, obtaining ancient ice cores so that he could measure lead contamination over time. Ice forms in layers from winter snow and can be dated by simply counting the seasonal layers. This revealed the onset of lead smelting by the ancient Greeks and Romans.

---

[37] 'Clair C. Patterson', interviewed by S.K. Cohen 1995, Archives of the California Institute of Technology.

Some historians believe that high exposure to lead was a contributing factor in the decline of the Roman empire! He found that the lead content of the ice rose dramatically beginning in 1923 when leaded petrol was introduced, rising to a thousand times as high as prehistoric levels.

The leading 'industry expert' was medical doctor Robert Kehoe. His research laboratory was generously funded by the oil companies, and his mission was to prove that exposure to low levels of lead was safe and to exonerate the petroleum companies from blame for any deaths and brain damage that it may have caused. Kehoe argued that Patterson's results were nonsense and that the amounts of lead found in the human body were natural. But Patterson was persistent and warned senators and congress about the dangers of lead pollution from automobiles[38]. He also analysed the bones and teeth of prehistoric people and found that humans were contaminated by atmospheric lead with an amount 100's of times that of our ancestors.

Today we know that the only safe level of lead in our bodies is zero lead. Ancient Romans were aware it could cause madness and death, though that didn't stop them from using it in pipes, cosmetics and as a wine sweetener. It was widely known by the 19th century that people working in the lead industry would develop sickness. But the extent of lead in the atmosphere and our bodies was not known until the work of Patterson. His research led to the phasing out of lead in gasoline, paint and plumbing. Thanks to false arguments and court challenges from the lead and petrol industries, this process was slow. Many countries had banned leaded petrol

---

[38] 'Clair Cameron Patterson', G. R. Tilton, 1998, Biographical Memoirs of the National Academy of Sciences, Vol. 74, 267-283

by the 1990s, but some countries continued to use it. The last country to stop using lead in petrol was Algeria in 2021.

In 1995 Patterson was awarded the Tyler prize for environmental achievement. Later that year he died from a severe asthma attack. It has been recently estimated that past exposure to lead directly causes hundreds of thousands of deaths each year, even today. Thanks to the work of Patterson, millions of lives will ultimately be saved.

20. Allan Sandage (1926-2010)

"for measuring the age of the galaxy and the universe"

We do not have the technology to extract any material from the Sun and bring it home to analyse in the laboratory. Even if we could, it is not obvious how we could use it to accurately measure its age. So how can we determine the age of stars like our Sun, things we cannot touch? What about the age of the universe itself? It was the astronomer Allan Sandage who first determined the ages of stars in our galaxy and narrowed down the precise age of our universe.

Allan Sandage was born in Iowa, America in 1926. Like many astronomers, his interest in astronomy happened at an early age when he first looked through a telescope of a childhood friend [39]. He was immediately determined to become an astronomer and set out on his path to achieve that goal. As a young researcher he worked with the well-known astronomers Walter Baade and Edwin Hubble, impressing them so much that he was offered a permanent position at the Mount Wilson Observatory even before he had finished his PhD. Sandage became one of the world's most influential astronomers from 1950 until his death in 2010. Over his career he spent over 2,000 nights at the telescope, the dome resonating to the sound of Richard Wagner – Sandage enjoyed opera.

By the 1950s, astrophysicists had worked out most of the details about how stars evolve over time. The time was ripe for Sandage to apply these models to the stars in our galaxy. Stars live and die by well-understood physical principles. They are much simpler to understand than how a simple blade of grass lives and dies. We can write down relatively simple equations to describe how they function and to predict

---

[39] 'Allen Rex Sandage', D. L. Bell & F. Schweizer, 2012, Biographical Memoirs of Fellows of the Royal Society 58, 245–264

how they evolve through time. We can approximate a star by a spherical ball of ionised gas or plasma. Its structure is an intricate balance between gravity (which always pulls material inwards) and the energy from fusion that prevents the star from collapsing into a much denser configuration such as a black hole.

A star is not a uniform sphere though; its density, temperature and chemical composition vary with radius and over time. Stars shine with a range of brightness and colours which depend mainly on their mass and age. The colour of a star is related to the temperature on its visible surface. More massive stars appear bluer to our eyes – they generate higher pressures and densities at their cores and they burn through their fuel rapidly and can have lifetimes of just a million years. Small stars appear redder since they are cooler and live far longer, over a trillion years. Just by measuring a star's surface temperature, its luminosity and its mass is enough to estimate how long it has existed as a star. We can also predict how it will evolve into the future and when it will eventually stop shining. Even stars do not last forever. Eventually, all stars will consume their nuclear fuel and die, some exploding as spectacular supernovae and others simply fading to darkness.

At the time Sandage was carrying out his PhD thesis, it was thought that red giant stars were stars at the beginning of their lives. He showed that the opposite was true – stars like our Sun end their lives as red giants after they briefly expand to an enormous size. Armed with all of this knowledge, Sandage set out to find the oldest stars in our galaxy to determine a lower limit to the galaxy's age. All we need to know is that the lifetime of a star depends mainly on its mass. A star that is twice as massive as the Sun will live for only 800 million years. Conversely, one that is half the mass will last for 20 billion years. How do we use this knowledge to estimate the age of our galaxy? Well, it turns out that stars often form together in large groups or clusters, such as the beautiful Pleiades star cluster often called the Seven Sisters[40].

Within star clusters the stars all form together at the same time but with a range of different masses. Depending on the age at which the stellar cluster forms, all the stars above a certain mass will have died and faded out of sight. This is how Sandage first determined a lower limit to the age of our galaxy. The most massive stars found in the oldest star clusters of our galaxy are measured to have about 70 percent of the Sun's mass, which means that these stars are over 10 billion years old! Parts of our galaxy are therefore nearly three times as old the Sun and Earth and the universe beyond our galaxy must therefore be at least this old.

Sandage then turned his attention to something grander, the age of the universe.

The rate at which space is expanding is called the Hubble constant, even though it was first measured by Lemaître, and it is actually not a constant as it changes with time. It is the parameter that relates the distance of a galaxy to its apparent speed away from us. To calculate the age of the universe, you

---

[40] Most people can see at most six stars of Pleiades with the naked eye, although the cluster actually contains hundreds of fainter stars. It is a mystery why most early cultures name seven stars. One theory is that the stories of its name were passed down for over 100,000 years, a time when the seventh brightest star was clearly visible. Over time the seventh star has moved very close to one of the other stars and our eyes resolve the two stars as just one. Take a look one clear night – it is a good test of your eyesight. Pleiades is the brightest visible star cluster. You can see it in the constellation of Taurus, close to Orion (the giant hunter holding a club).

need to know the value of Hubble's constant as it tells us how fast the universe is expanding. The basic idea is to calculate how long it has taken since the big bang for the galaxies to expand into their current configuration. When this was first calculated, the age of our universe was found to be just 1.8 billion years old, much younger than the age of the Earth and the stars!

It was only in 1952 that it was realised that the distance scale had an error and the measured value of Hubble's constant was completely wrong. Hubble's constant relied on cepheid-variable stars to measure distances to galaxies using 'Leavitt's law'. It was the German astronomer Walter Baade, Sandage's thesis advisor, who showed that the original cepheid-variable distances used to determine distances in our galaxy used a different class of stars from that which Hubble used to measure the distances to galaxies like Andromeda. Baade discovered that there are two basic types of stars. The so-called population I stars, like our Sun, which are located throughout the disk of our galaxy. They formed later in the history of our universe and a few percent of their mass comes from elements heavier than helium that formed in early generations of stars. The so-called population II stars are ancient stars, that formed early in the history of the universe and they contain a tiny fraction of the heavier elements. Population II stars are found in ancient star clusters called globular clusters and these were initially used to calibrate Leavitt's law. Hubble had then unknowingly used the same law on population I stars in galaxies, but its calibration was incorrect.

Baade's work in 1952 halved the Hubble constant and doubled the age of the universe. But this was still younger than the age of the solar system measured by Patterson and the ages of the oldest stars measured by Sandage. Soon after, Sandage discovered that Hubble had mistakenly identified ionised clouds of hydrogen gas as being bright cepheid stars. The brightness of these objects are certainly not standard candles. Correcting for these effects, in 1958 he published the value of 75 kilometres per second per megaparsec, a factor of seven smaller than originally estimated by Lemaitre and Hubble and close to the value found today[41].

Sandage single-handedly began a large observing program to measure the distances and velocities of a large number of galaxies in order to pin down the correct value of the Hubble constant. In 1961 Sandage wrote a paper titled 'The Ability of the 200-inch Telescope to Discriminate between Selected World Models'. It became the foundation of modern observational cosmology and turned cosmology into a quantitative precision science.

Thanks to the new observations of Sandage, by the 1960s the age of the universe using Hubble's constant was revised to be between 6 and 13 billion years – consistent with the age of the Earth, the solar system and the oldest stars. At this time, the large range of possible ages was due to the uncertainty in the composition of our universe and uncertainties in the measurement of the Hubble constant which still varied by a factor of two among different observers. This motivated the construction of the Hubble Space Telescope which could measure this parameter far more accurately.

During the 1980s Sandage became convinced of the existence of a creator and announced that he had become a christian. He thought that the cosmos and life were too

---

[41] 'H: The incredible shrinking constant 1925-1975', V. Trimble, 1996, Publications of the Astronomical Society of the Pacific, 108, 1073-1082

complex to be explained by science alone and that his research was providing the date of creation. He wrote that there is no conflict between science and religion if it is understood that each treats a different aspect of reality. That science tries to answer 'what, when and how?' but only religion could answer 'why?' However, his lifelong colleague, the Swiss astronomer Gustav Tammann wrote in his obituary of Sandage that *"In the end he highly valued Christian philosophy, but he did not find faith"*.

During his career, Sandage published over 500 research papers. According to his closest colleague Tammann, Sandage thought that work was the only meaningful human activity – that life was not about having fun. He died of cancer aged 84 in 2010 whilst still writing scientific papers, publishing over 40 in the last decade of his life.

21. Fred Hoyle (1915-2001)

"for the theory that elements are forged inside stars"

You may have heard the phrase 'we are stardust', that most of the elements in our bodies were made inside stars. This profound discovery is thanks to the work of English astrophysicist Fred Hoyle. Whereas Arthur Eddington had speculated that elements could be made in stars, Hoyle showed how it could be done.

Hoyle was born in Yorkshire, England into a working-class family, which was unusual at the time when most scientists came from privileged homes. His father, a wool merchant, joined the Machine Gun Corps during World War I, the period he was born during which time his mother earned money playing piano for the silent films in the local cinema. At school he had conflicts with his teachers and found the lessons boring. He often feigned illness so he could stay at home, where he studied a textbook in chemistry and carried out experiments with equipment he found in the house. As recounted in his 1994 autobiography, 'Home Is Where the Wind Blows: Chapters from a Cosmologist's life', in his parents' absence he enjoyed making gunpowder and creating explosions. Since school was compulsory, and his family were not wealthy, he did well to pass the countrywide scholarship exam such that he could attend grammar school where he began to excel. He was particularly motivated after reading a copy of Arthur Eddington's popular science book 'Stars and Atoms' at the age of 12. He would go on to win a scholarship to attend Cambridge University where he would be taught general relativity by Eddington himself.

Whilst at Cambridge university, he was awarded the Mayhew Prize for the best student in applied mathematics. He asked the famous physicist Paul Dirac if he could be his graduate student stating he would not need supervision since Dirac was known to dislike advising students and Hoyle was happy to work on his own. Hoyle's PhD thesis was on quantum electrodynamics but his interest in stars took him to astrophysics. In one of his first publications, he wrote about the possible effects of our Sun passing through an interstellar cloud, changing its radiation output and causing ice ages. Later, he would write a dozen books on science fiction, often bringing ideas from his research into his writing. I can recommend his first book written in 1957 'The black cloud', in which he describes the Sun becoming engulfed by a giant cloud of gas that seems likely to destroy most of the life on Earth by blocking the Sun's radiation.

Hoyle's research was interrupted by World War II during which he worked for the British navy developing radar systems. At the time radar was only used to infer the direction of an incoming aircraft but not its height. He realised that there was interference between the radio beam reflected from the aircraft and from the fainter indirect beam that reflected off the sea. He developed a technique by which this could be used to infer the height of the aircraft. Hoyle later wrote *"War would change everything. It would destroy my comparative affluence, it would swallow my best creative period, just as I was finding my feet in research."* However, as part of his work with the navy he was able to travel to America where he learned about the atomic bomb project and started to think about nuclear processes that could take place inside stars. This would lead to his discovery of the origin of the elements.

After the war Hoyle returned to Cambridge and began working on the evolution of stars. In 1946 he showed how the interior of massive stars reached temperatures of billions of degrees as they evolved. And that at the end of their lives these stars collapsed and exploded, scattering any elements that they might synthesise into the galaxy. But at that time astrophysicists could not see how elements beyond helium could be made. We have heard about how Gamow and Alpher could synthesise hydrogen and helium in the big bang, but the origin of the heavier elements still remained a mystery. The reason that heavier elements are not created during the big bang is that in stars it takes tens of thousands of years to convert a significant amount of helium to carbon. However, the conditions during the early universe only resembled the inside of a star for a matter of a few minutes.

By 1950 it seemed impossible that even stars could create any elements heavier than helium. Fusing hydrogen with helium would produce lithium-5 which is unstable. Fusing two helium nuclei would create beryllium-8 which is also unstable and decays back to lighter nuclei in less than 10^{-16} seconds. If during this time a third helium nuclei fused with the beryllium it could form a carbon-12 nucleus, however, the chance of this happening was thought to be extremely small. That is until Hoyle speculated that carbon-12 possessed an energy level very close to the combined energies of the beryllium and helium. This would lead to a resonant reaction with a much higher probability of happening. Soon after, laboratory experiments confirmed Hoyle's prediction!

Confirmation of the idea that stars create heavier elements came in 1952, when the astronomer Paul Merrill detected the presence of the element technetium in red giant stars. Technetium is the lightest element that has no stable isotopes. It decays into lighter atoms with a half-life of 4.2 million years. But the red giant stars in which it was found are billions of years old. The conclusion from these observations was that the technetium must be created inside of these stars.

By 1954 Hoyle had worked out many of the details of the synthesis of the elements up to nickel, which occur in a series of thermonuclear reactions in which the ashes of one stage become the fuel for the next – hydrogen fuses to form helium, helium to carbon, carbon to oxygen and so on. Nickel is the heaviest element that stars can make in this process. Fusing together elements lighter than iron and nickel produces more energy than that needed for the reaction. But that gain of energy which enables self-fuelling reactions to continue stops at nickel since heavier elements need more energy to fuse together than they produce from fusion. The origin of the heavier elements still remained unknown.

In work published in 1955 with the astrophysicist Martin Schwarzschild, Hoyle solved the equations by hand that describe the evolution of stars to show how stars like our Sun evolve into red-giant stars. Red giant stars are so luminous that you can see many with your eye, Arcturus, in the constellation Boötes, is one of the brightest stars in the night sky. Once all the hydrogen has been converted to helium at the core of our Sun, this process will continue in the outer regions of our star causing its outer envelope to expand so much that it will reach Earth's orbit. This will happen around five billion years from now. The core of our Sun will increase in density and temperature, allowing the fusion of three helium nuclei into carbon. Towards the end of its life, carbon can fuse with helium to form oxygen, and some oxygen can fuse with helium to form neon. These processes produce free neutrons, which can also be captured by the atomic nuclei to give rise to even heavier elements in the period table.

At this time the best calculating tool was the mechanical pinwheel calculator that could add, subtract, multiply and divide. Hoyle used the German built Brunsviga 10 device, but his and all previous calculations of stellar evolution were basically carried out with pen and paper. In 1956 he was the first astrophysicist to use the British EDSAC (Electronic delay storage automatic calculator). This was an early vacuum tube electronic computer inspired by John von Neumann's research. The machine filled a large room, needing eleven thousand watts to power 3000 vacuum tubes that function as modern transistor logic gates. It would be programmed by feeding it a long paper tape filled with punched holes. These early computers could do several hundred calculations per second which began to revolutionise many areas of research. Hoyle realised that such machines could revolutionise astrophysics and futuristic powerful computers featured in many of his science fiction novels.

Hoyle then began a collaboration with the American nuclear physicist and astrophysicist Willy Fowler. Together they published physical interpretations of the two main classes of supernova that will be described in the next chapter on Fritz Zwicky. This in itself was quite an accomplishment. But their most famous work was a lengthy 1957 paper they published together with Margaret and Geoffrey Burbidge titled 'Synthesis of the elements in stars', which combined theory with observations to show how elements heavier than iron and nickel could be synthesised within massive stars by a process in which existing nuclei can capture free neutrons. In this lengthy paper, Hoyle was testing his own theories, using laboratory experiments on nuclear reactions carried out by Fowler, and observational measurements of the elemental abundances made by the Burbidges.

In a second major research thread of his career, Hoyle worked on alternatives to the big bang model. As Lemaître's ideas became widely known, it seemed as if there were no other way of explaining the observed expansion velocities of galaxies. But the idea of a beginning of our universe did not appeal to many and alternative ideas were explored that retained the old ideas that the universe was timeless and static. Some scientists, such as Zwicky, argued that redshifts were not due to expansion, but to a time varying speed of light, or that photons lost energy as they moved through space. During the late 1940s cosmologists developed an alternative model of our universe called the steady-state theory, in which the universe had no beginning and will have no end. One of its main proponents was Hoyle, who actually came up with the name big bang during a BBC radio interview in 1949 in which he said such a theory was irrational. He made fun of Lemaître calling him *"the big bang man"*. The term is unfortunate, since it gives the impression that the universe began with an explosion from a point rather than an expansion at all points – a subtle but very important difference.

Hoyle and his collaborators developed an alternative cosmological principle to the Copernican Principle we use today. This was that the large-scale appearance of the universe was the same in any place *and at any time*. The key feature of the steady state model was an eternally expanding universe, infinite in age. To explain why the expanding universe did not change in time and why all the galaxies did not expand away out of sight, they postulated the continual creation of new matter. This new matter would form new galaxies in between the old galaxies. How this matter was created could not be answered and it also violates the fundamental principle of conservation of energy and mass. However, these objections were insufficient for it to be discarded since they could be applied to the big bang model which also does not explain why there is something and not nothing.

Hoyle disliked the big bang model for the fact it gave the universe a 'miraculous beginning' which gave room for a supernatural explanation of its origin. Indeed, in 1951 Pope Pius XII concluded that modern cosmologists had arrived at the same truth that theologians had known for more than a thousand years! In 1967, the famous physicist Steven Weinberg, stated that *"The steady state theory is philosophically the most attractive theory because it least resembles the account given in Genesis. It is a pity that the steady state theory is contradicted by experiment."* It was the detection of the microwave background that eventually excluded Hoyle's steady state universe.

Hoyle believed that research with lasting value comes from following ideas thought most unlikely by others. I heard Hoyle at scientific conferences still argue for the veracity of his steady state universe long after the microwave background had proven it untenable. He was certainly a controversial character but at the same time one of the most creative thinkers from 20th century astrophysics. For a more complete review of his life and his research, I can recommend the tribute by his collaborator Margaret Burbidge[42].

Hoyle's achievements led him to the position of Plumian professor of astronomy in Cambridge from 1958 until 1972, a position held previously by his mentor Arthur Eddington. He established a nationally funded institute for theoretical astronomy at the university of Cambridge in 1967. This was a great success, but getting the institute funded and operational was a struggle for Hoyle. Continued conflicts with his colleagues at Cambridge over the future of the institute led him to resign his position in 1972 – the same year he was

---

[42] "Sir Fred Hoyle", E. Margaret Burbidge, Proceedings of the American Philosophical Society, 2002, Vol. 147, No. 4, 405-412.

knighted by the queen. As he stated in a letter to a colleague, he wished to waste no more time on political infighting that was draining his energy and time.

Hoyle moved to the Lake District where he continued to do research in private. Isolated from the scientific community his writings became more controversial, on topics as disparate as Stonehenge, Darwinism and viruses from space. In 1974 Hoyle and his colleague Chandra Wickramasinghe developed the idea of panspermia – that life could originate in space and spread to other worlds on comets and asteroids. Although ridiculed at the time by biologists, we now know that many of the building blocks of life can indeed form in space and have been found within meteorites and interstellar clouds.

In his autobiography he recounts his love of the mountains and how he was proud of climbing all the Munros – peaks over 3000 feet in Great Britain. In 1997 Hoyle fell into a ravine in Yorkshire whilst hiking. He was in hospital for several months due to hyperthermia and broken bones. In the following years his health began to slowly decline, and he died in 2001.

In 1983 the Nobel Prize in physics was awarded to two scientists: the Indian astrophysicist Subrahmanyan Chandrasekhar for his work on the structure and evolution of stars. The second awardee was one of Hoyle's collaborators, Willy Fowler, who received the prize *"for his theoretical and experimental studies of the nuclear reactions of importance in the formation of the chemical elements in the universe."* Why Hoyle who pioneered this work and later developed the theory with Fowler, was not awarded a Nobel Prize, is a mystery that puzzles astronomers to this day. Some think the reason is that Hoyle, born in Yorkshire, England, was a blunt and opinionated scientist who had managed to upset most of his colleagues. However, his legacy remains – Hoyle was the scientist who showed how the elements could be produced, that we are made of stardust, and his lack of a Nobel Prize was one of the greatest oversights of the Nobel committee.

22. Fritz Zwicky (1898-1974)

"for discovering dark matter in galaxy clusters, interpreting supernova and predicting the existence of neutron stars"

Of all the unrecognised scientists of the 20th century whose research advanced our understanding of the cosmos, few would rank above the unconventional and visionary Swiss scientist Fritz Zwicky.

Zwicky was born in Bulgaria; his father was a Swiss businessman and his mother was from Czechoslovakia. His parents desired him to have a Swiss upbringing and when he was six years old, he was sent to live with his grandparents in the Zwicky ancestral home in Glarus, a peaceful mountain filled canton of Switzerland. At school his friends said that he excelled without trying and often helped others who were not as gifted. After being tested in 14 subjects at one of the toughest middle schools in Switzerland (Zurich Technical College, Industrieschule), he scored 82.5 out of 84 – the highest grade in the 150 years of the school's history[43].

He studied engineering, mathematics and physics at the Swiss Federal Institute of Technology and carried out his diploma with the brilliant theoretical physicist Herman Weyl. In 1922 he received his doctorate in theoretical physics, applying the new theories of quantum mechanics to the structure of crystals. His supervisor was Peter Debye who would go on to receive the Nobel Prize in chemistry in 1936.

Zwicky was offered a fellowship from the Rockefeller Foundation in America, a program to bring European scientists to America, particularly those who understood the new field of quantum mechanics developed in Europe. It was established by Robert Millikan, who was awarded the 1923 physics Nobel Prize for his experimental determination of the charge of an electron. Zwicky asked to be placed anywhere close to mountains and he was sent to the California Institute

---

[43] 'Fritz Zwicky and the search for dark matter', K. Winkler, Swiss American Historical Society Review, 50(2), 23-41

of Technology to where Millikan had just moved. Zwicky was a passionate skier and mountaineer and he carried out several first ascents in Switzerland, often descending by skiing down the glaciers. When he arrived in Pasadena he asked Millikan where the mountains were. After Mount Wilson was pointed out, Zwicky disappointedly responded that this was a mere foothill[44].

After impressing Millikan, Zwicky was offered an assistant professorship at Caltech in 1927. The following year he was offered an associate professorship at the Institute for Theoretical Physics at the University of Zurich, but he had just accepted the offer from Caltech. Caltech also had the world's largest telescopes – Mount Wilson hosted the 100-inch Hooker Telescope that Edwin Hubble was using to determine the nature of nebulae and Zwicky's interests were turning to astronomy.

In 1929 Zwicky published one of his first papers in astrophysics, arguing that the redshift of the spectrum of light from galaxies was not due to the expansion of the universe but was an effect of gravitation causing the photons to lose energy. For the rest of his career, he never believed in the big bang which antagonised many of his colleagues. Zwicky was proven incorrect but it revealed his willingness to challenge orthodoxy even at an early stage of his career.

His most brilliant contributions to astronomy came in the 1930s and were so far ahead of their time they were mostly ignored by the scientific community for several decades. In 1931 the German astronomer Walter Baade moved to America and began working at Mount Wilson Observatory. Zwicky

---

[44] 'Idea Man', S.M. Maurer, 2001, SLAC Beam Line, 31N1, 21

and Baade began working on the nature of the mysterious 'novae'.

For several millennia, astronomers had observed the appearance and subsequent disappearance of new stars in the night sky which they called nova. Before the invention of the telescope, about eight of these events had been witnessed. We have already heard about one of those, that observed by Tycho Brahe in 1572. It was Brahe that gave these events their name. By the 1920s astronomers with their telescopes were finding a dozen or more such nova appearing in the sky each year, and a similar number in the nearby Andromeda nebula. But a new class of nova was emerging, visible in very distant nebula that must be far more luminous than those in our Milky Way.

In 1933 Zwicky announced their theory at a conference and they published their idea the following year. They concluded that there must be two types of novae. The rare events, observed over the preceding 2000 years, they called 'super-novae' but today we write supernovae. They shone for a few weeks with a luminosity that they estimated to be comparable to an entire galaxy of stars. They correctly proposed that these events must arise from the explosion of stars at the end of their lives. Zwicky persuaded Caltech to purchase a special 'Schmidt' telescope that could be used to photograph several square degrees of the night sky. Over the next few years, he took hundreds of photographs of galaxies to search for the rare supernovae. He announced the discovery of several new supernovae in 1938 from which he concluded that a typical galaxy might only have one such event every few centuries.

There are two main ways in which a supernova can occur. In the most massive stars, above eight times the mass of our Sun, the gravitational squeezing is so intense that their core temperatures exceed 600 million degrees. At this temperature the carbon and oxygen atoms begin to fuse together to create shells of heavier elements all the way up to iron. The internal structure of the star resembles an onion, with an iron and nickel core surrounded by a layer of silicon, then a shell of carbon and oxygen surrounded by the outer layers of helium and hydrogen. Stars cannot fuse together nickel to form heavier elements though, since that would require more energy than is released from the fusion process and there isn't enough energy available. At this point the hot core of the star begins to cool. It cannot withstand the enormous gravitational forces and rapidly collapses – a violent implosion that once under way cannot be stopped. The massive iron core collapses at a speed of one hundred thousand kilometres per second, instantly creating temperatures of 100 billion Kelvin.

A large fraction of the energy from the gravitational collapse is transformed into neutrino-antineutrino pairs. About ten percent of the mass of the star, equivalent to 10^{46} joules of energy, is converted into neutrinos which stream outwards at the speed of light. It is a giant neutrino powered explosion. And all of this happens in just a few seconds. The gravitational collapse of the massive metallic core releases enough energy so that the expanding outer shells of matter shine as brightly as an entire galaxy of stars for several days. Such explosions are called Type II supernovae. The actual physics of the explosion is still an active research area; we think that it is driven by the intense flux of neutrinos produced, but the details are not yet fully understood.

The second way a star can explode is via the transfer of mass from one star to another. Most stars form with a companion, in a so-called binary system. This is thought to be a result of the fragmentation of the central core of collapsing gas within the star-forming molecular cloud. Some of these binary pairs are so close that the surfaces of the stars are actually touching. Others can be widely separated with orbital periods of many centuries. Because the stars are so far away, we cannot resolve pairs of stars with our eyes, they appear as just one star. In some cases, the binary orbit is oriented such that one star passes in front of its partner and their brightness suddenly dips. These are called eclipsing binaries and you can see some of these with your eyes. Look for the bright star Algol in the constellation Perseus. Its apparent brightness dips three times fainter about every three days, as its brightest small star is eclipsed by its larger dim companion.

If the binary stars are close together there can be a mass transfer from one to the other. This is happening right now to the star Algol. If one of those binary stars has already evolved into a white dwarf, it might reach a critical mass, 1.4 times the mass of our Sun. At this point it can reignite nuclear fusion which triggers a violent explosion called a Type I supernova. About one in five supernovae are of this type. Within a few seconds, a large fraction of the carbon and oxygen in the white dwarf undergoes runaway nuclear fusion, mostly into nickel-56 releasing 10^{44} joules of energy. Because there is a fixed critical mass at which a white dwarf explodes, their resulting peak luminosities are always similar – this allows their use as standard candles. Nickel-56 is unstable and decays into iron with a half-life of six days. Over half of the mass of the white dwarf is transformed into iron, but it also produces a significant fraction of the elements between silicon and scandium, and most of the abundance of the elements between titanium and zinc in the periodic table. The explosion ejects these elements into the surrounding interstellar medium – the iron in the Earth and within our bodies primarily originated from exploding white dwarf stars.

Zwicky persuaded George Hale, the director of Mount Wilson Observatory, to build a larger 48-inch Schmidt telescope at Mount Palomar. Its primary purpose was to photograph the entire northern sky and the resulting Palomar Observatory Sky Survey became a major cornerstone of astronomy for the next fifty years. But Zwicky also used the telescope to search for more supernova and he eventually discovered over 100, more than any astronomer, until the large automated supernova searchers at the end of the 20th century. With this telescope he produced a six-volume catalogue of 30,000 galaxies which was an invaluable source for astronomers until it was superseded by large galaxy surveys in the 1980s.

There was more. In their 1934 paper, Baade and Zwicky correctly speculated that supernovae could be responsible for generating the puzzling cosmic rays – high energy particles which appeared to be colliding with the Earth from unknown sources. Cosmic rays are charged particles, such as protons and helium nuclei, which are thought to gain their energies from inside the expanding shock fronts of supernovae remnants. When a star explodes, its debris expands into the interstellar medium close to the speed of light and carries a magnetic field which acts like a giant cosmic particle accelerator. The highest-energy cosmic rays are travelling very close to the speed of light and have energies that are up to 10 million times higher than the protons inside the Large Hadron Collider beam tunnel at CERN! That is quite something: A single subatomic particle with an energy equivalent to the record-breaking 161-kilometre-per-hour cricket ball bowled by Shoaib Akhtar of Pakistan against England during the 2003 World Cup. If that single proton had collided with you, it might have knocked you over! Cosmic rays fill our galaxy and are continuously striking the Earth at a rate of about one per square centimetre each second. Luckily, our atmosphere does a remarkably good job at shielding us from these particles – the cosmic rays collide with the molecules of air which prevents most from reaching the ground. Cosmic rays are a serious threat to space travel since they can damage the DNA in our cells and they can destroy sensitive electronic devices.

Zwicky and Baade also postulated that following the supernova, material at the centre of the star could collapse to form an extremely dense core made of neutrons, formed by the crushing together of protons and electrons. They called this remnant of a dead star a 'neutron star'. This was a remarkable insight as neutrons had only been discovered in 1932. Although this idea was ignored by the astronomical community, by the end of the decade, several theoretical physicists had calculated the properties of such objects, showing that they should only exist in a certain range of masses. Zwicky certainly had the knowledge to have made these more detailed calculations, but this was an early indication that he was less interested in the details and more interested in the ideas.

We now know that when a star explodes as a supernova, if the star was less than 25 times the mass of our Sun then the material at its centre would collapse and form a new state of matter - a neutron star. You will soon hear the story of their discovery and learn more about their nature when I describe the unrewarded work of Jocelyn Bell. But even a neutron star is not the ultimate fate of the largest stars. For stars with a mass above 25 times the mass of our Sun, the central iron core is so massive that not even quantum mechanical effects can hold matter together. At the end if its life, the core of the star undergoes a violent collapse, neutron degeneracy pressure is overcome and the final structure becomes so dense that it curves spacetime right back on itself and even light cannot escape from its surface – a black hole.

Zwicky relied on intuition and was often criticised by colleagues who complained that he never rigorously followed up on his ideas. In this way he was rather like George Gamow, moving from topic to topic and along the way hitting upon some brilliant ideas. And just like Gamow, some of those ideas later proved to be plain wrong, others were pure genius.

Most astronomers today associate Zwicky with the discovery of dark matter in clusters of galaxies. Gravity relentlessly strives to pull matter together and is the force responsible for creating an entire hierarchy of structure in the universe, from individual stars to the giant clusters of hundreds of galaxies. Galaxy clusters are the largest systems to have emerged from the gravitationally driven structure-formation process. They are regions of the universe which contained enough matter to reverse the cosmic expansion, resulting in a gravitational collapse and the formation of an equilibrium configuration of galaxies. It was thanks to Zwicky's surveys with the Schmidt telescope that numerous such systems were discovered – Zwicky noticed that nearly all galaxies were part of such groups and clusters. He went on to make the first measurement of the mass of one of the largest known cosmic structures, the Coma galaxy cluster, and found a rather unexpected result – one that is still not explained to this day.

Zwicky found that the galaxies are whizzing about in the Coma cluster at thousands of kilometres per second, from which he derived the cluster's mass to be equivalent to 10^{15} suns. But, when he actually counted the galaxies and estimated the total number of stars, he found a number that was one hundred times smaller. There simply was not enough mass in the observed stars and galaxies to hold the cluster together. To prevent the galaxies in the cluster from dispersing and simply moving away from each other, he speculated that there must be an enormous amount of mass in the cluster that he could not see. He called it *dunkle Materie*, or dark matter, the name given to it by earlier researchers who had speculated on its existence. Later, it was found that clusters of galaxies also contain a large amount of hot gas, mainly hydrogen and helium. But even this extra material in between the galaxies would not provide enough matter to hold the clusters together.

Zwicky's observations were found intriguing by other scientists. However, his results were regarded as a curiosity that perhaps could be explained in other ways. Some astronomers argued that the clusters of galaxies he was observing were expanding thus giving rise to the high speeds of the galaxies. It was not until the 1970s that dark matter was taken seriously after it was discovered within galaxies including our own Milky Way.

Zwicky's interests and skills were very broad. Around 1941 the Hungarian-American physicist Theodore von Kármán established the Aerojet Engineering Corporation [45]. Fritz Zwicky became its director of research from 1943. Aerojet

---

[45] In his autobiography, Kármán discussed the imprecision in determining the boundary between Earth's atmosphere and space as the point in which aerodynamics stops and astronautics begins because of the lack of atmosphere to contribute to lift. He writes *"Below this line, space belongs to each country. Above this line there would be free space"*. He stated this occurs 57 miles high. Today the edge of space is often called the Kármán line and is defined as 62 miles, or 100km above Earth's surface.

rapidly became the world's largest manufacturer of rockets and propellants, playing a key role in the defence of the United States. Within twenty years it grew from a dozen people to a company with 34,000 employees and a billion-dollar turnover. Zwicky made many contributions to the design of jet engines and their fuel, securing dozens of patents for his work. During World War II he had to obtain a special security clearance to work on these projects since he was still a Swiss citizen. By 1955 he was told that his clearance would be revoked unless he took up American citizenship. Zwicky refused to take out naturalisation papers because he thought naturalised citizens were second class citizens as they could never become president and they could even have their citizenship revoked!

In one of his more novel ideas, Zwicky attempted to create artificial meteors by placing an explosive filled with steel balls in the nose of a V2 rocket. He thought that by observing the resulting artificial meteors from the Earth we could learn more about the nature of meteors and of Earth's atmosphere. Although the first attempt failed, he put that down to human failure and constructed the second experiment himself. It was successful! Moreover, he calculated that some of the objects would have acquired enough speed to leave Earth's gravitational pull and orbit the Sun. He proudly announced he was the first to place an artificial object into orbit around our star!

Unfortunately, his confrontational manner and outspoken nature with his colleagues probably cost him any chance of receiving the Nobel Prize and relegated his insights into relative obscurity. When he didn't like someone, he was happy to tell them that. According to people who knew Zwicky well, one of his favourite insults was to call them "*spherical bastards*", because, as he explained, they were bastards no matter which way one looked at them. In his 1971 book which catalogues many different types of galaxies he begins with an introduction that describes his disdain for many of his colleagues by name: *"Today's sycophants and plain thieves seem to be free, in American Astronomy in particular, to appropriate discoveries and inventions made by lone wolves and non-conformists,..."* and *"The naivety of some of the theoreticians, at all times, is really appalling. As a shining example of a most deluded individual we need only quote the high pope of American Astronomy, one Henry Norris Russell..."*[46]

Zwicky was a self-proclaimed genius who some of his colleagues called an egomaniac. His genius was certainly apparent. The renowned French astronomer Gérard De Vaucouleurs stated that *"Of the astronomers of the 20th century he is perhaps one of the very few, if not the only one, to whom the word 'genius' can apply. This one was a genius, and he was very impatient with those not quite up to his measure."*[47]

Despite the often-repeated stories of his attitude towards his colleagues, Zwicky was a humanist. He collected old scientific books and manuscripts and donated over one hundred tons to libraries that lost their collections during the second world war. He was involved for many years with the Pesalozzi Foundation, supporting orphaned children and received their Gold Medal in 1953 for his humanitarian work.

By the end of his career, Zwicky had published over 300 scientific articles, ten books and held several dozen patents. In 1949 he was awarded the Medal of Freedom by President

---

[46] Catalogue of Selected Compact Galaxies and of Post-Eruptive Galaxies, 1971, F. Zwicky, self-published book.

[47] Gerard De Vaucouleurs interviewed by Ronald Doel, 1991, Oral History Interviews, American Institute of Physics.

Truman for his work on rocket propulsion during World War II, the first foreigner to receive that honour. He was even nominated for a Nobel Prize in 1935. But it was only in 1972 that he received any recognition for his research, when he was awarded the Gold Medal of the Royal Astronomical Society two years before his death.

23. Vera Rubin (1928 – 2016)

"for the observations that convinced scientists that dark matter exists"

When Vera Rubin was twelve years old, she would stay up late at night and through her bedroom window she would watch the stars slowly move across the night sky[48]. She would look for meteors and make maps of their trails. By high school she knew that she wanted to become an astronomer. She thought it would be no problem, after all, she had read about Maria Mitchell, the 19th century American astronomer who had won a prize for discovering a comet in 1847. She pursued her dreams of becoming an astronomer, despite advice from her teachers that she should stay away from science or should consider becoming an artist[49]. Her teachers put her off physics so much that she only applied to universities where she could major in astronomy. She won a scholarship to Vassar College and later she would become the first woman allowed to observe at the famous Palomar Observatory and she would go on to discover that all galaxies are held together by dark matter.

Rubin studied for her Master's degree in astronomy at Cornell and was taught by the likes of Richard Feynman and Hans Bethe. For this work she studied the 108 galaxies with measured radial velocities in order to explore whether there was a systematic motion of galaxies in addition to the universal expansion of space between them. Her results were rejected for publication by the two main astronomy journals. Decades later, in the 1980s the motions of galaxies would become a major research theme in astronomy.

When she tried to enrol in a graduate astronomy program at Princeton in 1948, she was turned down because of her

---

[48] 'An interesting voyage', V. Rubin 2011, Annual Review of Astronomy and Astrophysics, 49, 1-28

[49] 'How Vera Rubin encountered dark matter', J. Mitton, 2021, Astronomy & Astrophysics, vol. 62, 2-28

gender[50]. After she graduated from Cornell, Rubin (then Vera Cooper), married Robert Rubin who was a graduate student in chemistry. Rubin's husband began to work at the Johns Hopkins Applied Physics Laboratory and shared an office with Ralph Alpher. George Gamow heard of her results on the motion of galaxies and called her to learn more and later would become her PhD supervisor. Rubin completed her PhD thesis in 1954 exploring the distribution of galaxies in space. Despite limited data at the time, she found that galaxies were not distributed randomly in space – a result that would become a major focus of research in cosmology two decades later. When she sent her work to be published in the Astrophysical Journal, the chief editor Subrahmanyan Chandrasekhar declined to publish the paper saying that a student of his own was working on the same subject so she should wait until his work was done![40]

For the next decade Rubin juggled a dual family career at a time when married women were expected to stay at home. She brought up four children who would later all become scientists, whilst she worked in various temporary jobs in a role typically assisting other astronomers. During this time Rubin became interested in the dynamics of galaxies and wondered where a galaxy ended and where intergalactic space began.

For over half a century, astronomers had wondered if there were more to the galaxy than one could see. Lord Kelvin (William Thomson) stated that there could be unseen matter in the form of faint or dead stars. He even attempted to measure its presence using the motions of nearby stars. Henri Poincaré was the first to use the term 'dark matter' (matière obscure)

---

[50] 'Vera Rubin (1928-2016)' N. A. Bahcall, Nature obituaries, 542, 32

when discussing the work of Kelvin. From the early 20th century astronomers began to measure the rotational patterns of nearby galaxies, in particular the Andromeda nebula. By 1920 it was known that galaxies rotated even before it was known what galaxies were.

Most of the stars and gas in spiral galaxies like our Milky Way live in a flattened rotating disc-like structure. Our star, the Sun, takes around 250 million years to complete its orbit around the centre of our galaxy. One 'solar birthday' ago, the dinosaurs were first appearing on Earth. Two orbits ago, the first creatures began to explore the land on Earth. 16 orbits ago, simple single celled life appeared on Earth – astronomical timescales indeed.

In the same way that the mass of our Sun can be measured using Newtons law of gravitation, the Earth's orbital period and its distance from the Sun, we can calculate the mass of an entire galaxy by measuring the orbital speed of the stars or gas as far away from the centre of the galaxy as possible. Since most of the stars in a galaxy are concentrated within their inner regions, it was thought that the more distant stars should be moving more slowly than the inner stars, just like the planets orbiting the Sun; Mercury completes its orbit in just 88 days, whereas Neptune which is much further from the Sun takes over 60,000 days. The reason for this is that the gravitational force from the Sun is much weaker at the distance of Neptune.

By 1960 most astronomers believed that the rotation speeds of galaxies declined with distance. However, not all astronomers agreed that galaxies were so simple. By the 1930s there was some evidence that the Andromeda nebula might contain some missing mass. In 1940 the famous Dutch astronomer Jan Oort wrote of the galaxy NGC 3115 *"It may be concluded that the distribution of mass in the system must be considerably different from the distribution of light ... The strongly condensed luminous system appears embedded in a large more or less homogeneous mass of great density."*[51] But the observations by different astronomers were inconsistent and there was no consensus that any mass was missing.

In 1962 Rubin was teaching an evening class to six part-time graduate students in astronomy. Together they carried out a class project with the aim of studying the motions of stars in the Milky Way. They found that the orbital speeds of stars are all similar no matter how distant they are. Our galactic rotation curve was found to be flat beyond the distance of our Sun and did not decline as expected from the distribution of stars. The most plausible explanation is that there is considerable matter within our galaxy that could not be seen. Rubin and her students were the first to discover dark matter within our own galaxy. Following the publication Rubin recalled receiving many negative and unpleasant comments stating that her work could not be correct. Recognising her talents as an observational astronomer, Allan Sandage invited her to use the telescopes at Mount Palomar. Previously off-limits to women, she was the first to be allowed to use them. During her first observing trip, she drew a skirted woman and put it on the door of one of the toilets.

She obtained a part time position at Carnegie in 1965. It was there that she met Kent Ford who had just developed an image tube spectrograph – a sensitive device for measuring the spectrum of starlight. It is by measuring the small changes in spectral lines that the velocities of stars can be measured using the doppler effect. Rubin realised she could use this to

---

[51] J.H. Oort, 1940, Astrophysical Journal, 59, page 273.

answer her long-standing question about where galaxies end. Rubin and Ford began a collaboration that lasted several decades. They took data on many galaxies, looking beyond where astronomers had been able to look before. They found the surprising result that the speed at which the stars rotated about the centres of their galaxies was always constant with distance, nothing like the expected decline with distance. This result implied that all galaxies contained a large component of matter that was not readily visible – dark matter that stretched at least as far as the outermost stars to enable their high orbital speeds. The idea that galaxies were embedded within quasi-spherical distributions of dark matter that could be made of a new undiscovered particle was proposed in the 1970s.

Today we know a lot more about the quantity of dark matter and how it is distributed across the universe. There are at least ten different and independent ways in which we can detect its presence. One of the most powerful and convincing measurements can be made using gravitational lensing, which relies on the link between gravity and spacetime that was made by Einstein. Gravity literally curves the fabric of space, implying that photons of light do not follow simple straight lines through the universe. Every photon that we see from a remote star in a distant galaxy has travelled across the universe on its own roller-coaster journey to get here, curving around every massive object in its path. The deflection of light by massive objects is very similar to the bending of light by a magnifying glass; more highly curved lenses distort the images more strongly. In an analogous way, images of distant galaxies that happen to lie behind a galaxy cluster will be distorted. The amount of distortion can be used to measure the curvature of space due to gravity and hence the cluster's mass.

Dark matter has received that name because it is literally dark – perhaps made up of a vast number of unknown particles that do not emit light. Many thousands of research papers have been published on dark matter, and its existence, quantity and distribution in space are well established. I have written research papers on dark matter, many of my students have written papers on it and my students' students have written papers on it. However, despite much research its precise nature still remains a mystery. Many decades have passed since Zwicky, Rubin and their colleagues identified a component of matter that is unlike anything with which we are familiar. A host of different observations all lead to the same answer - that the atomic material that constitutes us, as well as planets and stars, is just a minor part of the universe; there is about five times as much dark matter as atomic matter. That is certainly another profound thought: even the atoms of which we are made are only a small fraction of all the matter that exists in the universe!

Rubin received many awards and prizes for her research, from the Gruber Cosmology Prize to the National Medal of Science. Many astronomers expected that she was going to be awarded the Nobel Prize, yet she never received that accolade. This was a missed opportunity that would have helped inspire future generations of women scientists. She was not the first to discover evidence for dark matter, nor to make observations of galactic rotation curves[52]. Amongst others, the work of Dutch radio astronomer Albert Bosma also deserves recognition. But it was Rubin's dedicated work over several decades collecting high quality data on many galaxies that convinced scientists that galaxy rotation curves were flat and that dark matter (or

---

[52] 'One hundred years of rotating galaxies', V. Rubin 2000, The Astronomical Society of the Pacific, 112, 747-750

perhaps more accurately, missing acceleration) existed. And there are certainly precedents in Nobel Prize awards for those scientists whose observations convince the community, as we have seen with the rediscovery of the cosmic microwave background, but also in the case of the 2019 Nobel Prize for the discovery of exoplanets about which we will hear later.

Vera Rubin died of natural causes in 2010. She inspired a generation of female scientists to pursue their dreams and she left them with these thoughts: in the preface of her 1997 book, 'Bright galaxies, dark matter', she writes "*We have peered into a new world and have seen that it is more mysterious and more complex than we had imagined. Still more mysteries of the universe remain hidden. Their discovery awaits the adventurous scientists of the future. I like it this way.*"

Missing acceleration? Vera Rubin was an observational astronomer and loved observing and discovering more about our universe. She never advocated what her results implied for the nature of the dark matter. What she found and what Fritz Zwicky found, was evidence for missing mass *if gravity behaved as Isaac Newton or Albert Einstein postulated*. What if all the dynamical effects of dark matter were due to Einstein and Newton being wrong – that gravity was more complicated? Anyone to suggest such an idea would run the risk of being ridiculed by the scientific community, yet the Romanian born scientist Mordehai Milgrom dared to propose such an idea in 1982.

Milgrom attempted to explain the rotation curves of galaxies, not with an excess of unseen matter, but by proposing that the law of gravitation changed from an inverse square force to one that decreased linearly with distance once accelerations became low. Milgrom noted that Newton's laws are well tested in regions where gravitational forces are high, such as in the solar system or on Earth. But they have not been tested where the forces are very weak, such as in the outskirts of galaxies. With his modification to Newton's law of gravitation, one can easily produce the flat rotation curves of galaxies without the need for any missing matter. Whilst Milgrom's theory was based on Newtonian physics, there have since been complete theories developed using relativistic physics that allow such theories to be tested on cosmological scales. Whether or not the dark matter is a missing particle, or a failure of our theories of gravity remains to be seen.

24. Rainer Sachs (1932-), Arthur Wolfe (1939-2014) and Joe Silk (1942-)

"for predicting temperature variations in the cosmic microwave background photons"

Following the discovery of the big bang photons, it was soon realised that if you could measure their properties in more detail, they would reveal a vast amount of information about the conditions in the early universe. This prediction and some of the first detailed calculations were first published in 1967 and 1968 by the German-American physicist Rainer Sachs, the American astrophysicist Arthur Wolfe, and by the British astrophysicist Joe Silk. Their pioneering work led to the launch of several space telescopes to observe the microwave background in more detail, culminating in the recent Planck satellite that nailed down our place in the universe and its physical structure to very high precision.

As the big bang model was being developed and tested, many scientists began to think about how galaxies form within an expanding universe. It was realised that all that was needed were some small irregularities in the early distribution of matter to act as seeds for the subsequent growth of structure. Such theories began to be developed in the 1930s by scientists such as Georges Lemaitre, George Gamow and Evgeny Liftshitz. Decades later, these models would be extended in great detail by Jim Peebles and Yakov Zeldovich about whom we will hear later.

In 1966 the Soviet cosmologist and Nobel Peace Prize winning dissident Andrei Sakharov suggested that such initial irregularities could be the result of quantum fluctuations during the early universe. From these initial fluctuations, entire galaxies could be formed by the attractive force of gravity acting to pull matter together from a large enough region of the universe. Sachs and Wolfe working together and Silk working alone, realised that the presence of these matter irregularities should reveal themselves as tiny variations in temperature of the cosmic microwave background photons.

Their work was extended and refined by many cosmologists, including 2019 Nobel Prize winner Jim Peebles. These variations in temperature would provide a very precise picture of conditions in the early universe and could be used to measure many of the cosmological parameters, such as the geometry and age of the universe, the matter and baryon densities and the nature of the primordial fluctuations.

Rainer Sachs was born in Frankfurt am Main in 1932. His father was a Soviet born metallurgist from a Jewish family. They fled Germany in 1937 to avoid Nazi persecution. Rainer went on to study mathematics at MIT and theoretical physics at Syracuse University. Until 1985 he worked on relativistic cosmology and astrophysics, before switching to mathematical and computational biology. Much of his research was pioneering and his name is associated with many technical, but important results, such as the Goldberg-Sachs theorem and the Ehlers-Geren-Sachs theorem.

Arthur Wolfe was born in New York and studied physics at university. He carried out his PhD supervised by Rainer Sachs in which he calculated the fluctuations in the cosmic microwave background fluctuations. He then changed career directions and became an observational astronomer and devoted his research to uncovering the development of structure in the universe over cosmic time. With his bushy moustache and booming voice, he was always a formidable presence at the conferences I attended.

Joe Silk studied mathematics at the University of Cambridge and earned his PhD in astronomy from Harvard in 1968. He spent most of his career at the University of Berkeley before returning to Oxford University in 1999. To work alongside Joe Silk was one of the reasons I took my first postdoctoral position in Berkeley in the early 1990s. I recall memorable dinner parties at his home high in the Berkeley Hills overlooking the Golden Gate Bridge – he would invite young researchers and PhD students and we would talk about science until the early hours over bottles of fine French wine. He would hand out unsolved problems like candy, giving us hints as to how we might tackle them. Silk is still highly active in research and has influenced research in astronomy, astrophysics, astro-particle physics and cosmology. I believe he is the most prolific scientist in these areas of research, having published as author or co-author over one thousand scientific articles as well as many books. Here I will focus only on his second and fifth publications, which like Wolfe's, were also part of his PhD thesis research.

In 1967 Sachs and Wolfe showed how the big bang photons lose energy and are gravitationally redshifted as they cross regions of the universe with a higher matter density. Independently in the same year, Silk showed how temperature variations in the microwave background from matter irregularities could persist through cosmic time. The following year he showed how the radiation can diffuse away from smaller fluctuations, which damps the fluctuations below a certain scale, leaving a characteristic imprint in the primordial photons.

These works inspired other cosmologists to calculate in great detail all of the physical processes that leave their imprint on the microwave photons. The calculations are complex, but they basically treat the evolution of the universe as an expanding fluid consisting of matter and photons. During the hot and dense phase of the early universe the photons, produced by the annihilation of matter and anti-matter, and the plasma, the protons and electrons, are coupled together and are constantly scattering off each other. Regions with a higher matter density gravitationally attract this fluid but this is resisted by the pressure of the photons. This causes sound waves or acoustic oscillations in the fluid. The photons and plasma undergo a series of oscillations of expansion and collapse within these regions. As it expands, the photon-baryon material cools down, as it contracts it heats up. As the microwave background photons travel across these regions of matter irregularities in the universe, they can gain or lose a small amount of energy. This creates a series of characteristic temperature variations which holds information about the properties of our universe.

Then, after the universe has expanded for about 400,000 years, the plasma cools sufficiently so that neutral atoms can form without being blasted apart by the most energetic photons, and this whole process comes to a halt. The photons are then free to stream in all directions but they retain this imprint of temperature variations from the early universe which we can observe today. The size of these temperature variations was predicted to be tiny – at the milli-Kelvin level – but this was almost within reach of technology in the 1980s. To detect them required developing very sensitive detectors that needed to be launched high above Earth's atmosphere or into space. The race was on.

In 1983 a Soviet satellite called RELIKT-1 was launched in the first attempt to measure the temperature variations of the microwave background. The researchers involved in the effort came tantalisingly close and even claimed to detect their presence, although the error bars were large. It was not until 1992 that NASA's COBE satellite achieved their unambiguous detection. After the discovery was announced, Stephen Hawking said in a Reuters news interview, *"It is the discovery of the century, if not of all time"*. The data from the COBE satellite confirmed the predictions of Sachs and Wolfe. But to extract more information, higher resolution observations were necessary.

Following COBE, dozens of experiments were carried out, including by sophisticated telescopes on the ground and carried by balloons, which confirmed the presence of 'Silk damping'. Then the WMAP (Wilkinson Microwave Anisotropy Probe) space satellite was launched in 2001. WMAP gave a detailed picture of the conditions that were present during those first few hundred thousand years and from which all structure in the universe evolved. This was followed by the European Space Agency's Planck mission, launched in 2009. The Planck satellite was designed to extract all the information possible from the microwave background and to measure the cosmological parameters of our universe at the highest precision possible. It has tested theories about the very early universe and the origin of cosmic structure. It has measured to a precision of about one percent, the cosmic inventory, the geometry of our universe, its expansion rate during the early big bang and much more. The final results were published in 2018, a 61-page summary of findings with 190 authors - a true landmark in precision cosmology that will be very difficult to supersede.

The importance of the cosmic microwave background in verifying the big bang cannot be understated. It was the detection of variations in temperature that gave us all the information we needed about the early universe. The Nobel Prize in physics in 2006 was awarded to the American astrophysicists George Smoot and John Mather for their role in the COBE satellite mission. But I think that the scientists who predicted their existence also deserve a large part of this recognition. Rainer Sachs and Arthur Wolfe never received any of the prizes or medals that they could have been awarded for their work. Joe Silk received the gold medal of the Royal Astronomical Society and the Gruber Prize for outstanding research in cosmology, both awards mentioning his work on developing the theory of cosmic microwave background fluctuations.

25. Marietta Blau (1894-1970)

"for developing the photographic method of studying nuclear processes and for discovering cosmic ray disintegration of atoms"

The Nobel Prize in Physics in 1950 was awarded to the British physicist Cecil Powell "for his development of the photographic method of studying nuclear processes and his discoveries regarding mesons made with this method." Powell began his research into his prize-winning work in 1938, however, the same techniques had already been developed over the previous decade by the Austrian physicist Marietta Blau. Powell learned everything he knew about the photographic method from Blau as well as their use in detecting high energy particles. She also applied them to the study of cosmic rays, discovering in 1937 how cosmic rays can break apart atoms. Blau was unrewarded for her research that also led to the final piece of the puzzle in the origin of the elements.

Marietta Blau was born in Vienna into a prosperous Jewish family. Her father founded one of the most prominent music publishing companies in Europe. She ranked top of her school class in maths and science and in 1914 she began to study physics at the University of Vienna and graduated with her PhD in 1919 aged just 24 on the topic of radiation physics. At this time, Austria had become a centre for research into radioactivity, thanks in part to its uranium mines in Bohemia.

In 1923 Blau took an unpaid temporary position at Vienna's Institute for Radium Research. Financially supported by her family, consultation for industry and other means, she began to investigate how specially formulated photographic emulsions could be used to capture the tracks of high energy subatomic particles. When the possibility of obtaining a permanent research position came up, she was apparently told that being a woman and also a Jew was too much[53]. However, at this time nearly one third of the institute's scientists were women, which was largely attributed to the influence of Marie Curie and the positive support of women by the director of the institute Stefan Meyer[54].

At the beginning of the 20th century, scientists could detect high energy particles using Geiger counters, and they could see the tracks the speeding particles left behind using a cloud chamber. This is a sealed container of gas and vapour and when a high energy particle passes through it it can knock the electrons from the gas molecules. These ionised particles then act as condensation points and a visible drop forms around them. But such devices could not record the tracks, so Blau turned to photography.

Blau experimented with different photographic emulsions and chemical mixtures, collaborating with the film manufacturers Ilford and Kodak. It took her many years to perfect the chemicals and thickness of the emulsions so that photographic film could be used as a particle detector. She persevered until the high energy particles left characteristic tracks in the emulsions that could be used to identify the particles and their energies. In 1937, together with her student Hertha Wambacher, she was awarded the prestigious Lieben Prize of the Austrian Academy of Sciences for the first detection of neutrons with this method. That same year she set out to record the mysterious 'cosmic rays' using the same techniques. Cosmic rays were discovered to be extra-terrestrial particles by the Austrian physicist Victor Hess some decades

---

[53] 'Marietta Blau: Between Nazis and Nuclei', P. L. Galison, 1997, Physics Today, vol 50, 42-47

[54] 'Marietta Blau: Pioneer of photographic nuclear emulsions and particle physics', R. L. Sime, 2013, Physics Perspectives, 15, 3-32.

before, whilst he was an assistant to Meyer at the Radium Institute.

Together with Wambacher, she placed stacks of the photographic plates at the science station on the 2300 metre Hafelekar peak near Innsbruck for a five-month exposure. When she examined the exposed plates, she found numerous tracks that were clearly made by high energy protons - cosmic rays. Some of the tracks were so long that the particles had energies higher than anyone had ever seen before. But most interestingly, some of the highest energy tracks led to a point at which they stopped and an explosion of new tracks seemed to begin, rather like a star-burst. Some of the new tracks were protons, others indicated alpha-particles that resemble the nuclei of helium. Blau correctly speculated that the highest energy cosmic rays were colliding with atoms of silver and bromine within the emulsion, shattering them into pieces. She had discovered what we now call cosmic-ray spallation.

Our bodies are made up from about 10^{28} different molecules which are themselves composed of individual atoms consisting by weight of 65 percent oxygen, 19 percent carbon, 10 percent hydrogen, 3 percent nitrogen and 3 percent everything else – over a dozen elements in trace amounts that are essential to life. Over half of our body weight is made up of water. The hydrogen nuclei in the water were made within the first second of the beginning of the universe. Thus, a large fraction of your body is 13.8 billion years old! The carbon and oxygen in our bodies were forged in stars like our Sun. And not just one star – they originated from literally thousands of different stars from all over the galaxy. By mass, 90 percent of your body is made of stardust, elements created in the nuclear furnaces at the centres of stars that lived and died before our own solar system even came into existence. Many of the essential trace elements that life uses, such as zinc or iodine, were made in the cosmic explosions from dying stars or even in the mergers that took place between neutron stars. Our bodies are indeed made of elements synthesised in stars, along with a heavy dose of hydrogen created during the first second of the big bang.

Except boron.

You many have noticed that when I was describing the production of elements in stars, I went directly from helium to carbon and skipped some elements, what about lithium, beryllium and boron? Boron in particular is essential for life on Earth. In animals it helps bone growth. In plants it is an essential element that is used for structural stability and cell division. Stars cannot synthesise these elements, since fusing hydrogen with helium would produce lithium-5 which is unstable – it decays into helium within a tiny fraction of a second. Fusing two helium nuclei would create beryllium-8 which is also unstable. In fact, all nuclei with a total number of protons and neutrons that equals 5 or 8 are unstable. So how do these elements arise on Earth?

The answer is due to cosmic rays – super high energy protons or atomic nuclei. These particles are accelerated close to the speed of light by extreme events in the cosmos, from exploding stars to the accretion of matter by supermassive black holes. They travel through the galaxy and even between galaxies. Some of these cosmic rays will collide with our Earth, hitting atoms of carbon and oxygen in our atmosphere. If the cosmic rays have enough energy, they will shatter these atoms into pieces and create lighter elements such as stable isotopes of boron. Plants and trees would perhaps not exist without supermassive black holes!

Thanks to the work of Marietta Blau who first showed how cosmic rays can break apart atoms, we now know how these light elements arise. Blau published her results in the journal Nature in 1937 and began to carry out experiments placing her photographic emulsions on balloons so as to make measurements of cosmic rays higher up in the atmosphere. However, in 1938 Blau was forced to flee Austria because of her Jewish descent and the antagonization of Nazi supporting colleagues within the Institute. Thanks to the help of Albert Einstein, she obtained a teaching position in Mexico where she spent several years unable to perform any research much to her distress. She appealed to Einstein to help her secure any research position on any topic, and he helped her move to the United States in 1944. In 1960, Blau returned to Austria to work at her old institute for four years, again without pay. Living her last years in poverty, subsisting off a meagre pension from her work in the United States, she died of cancer in 1970, most likely due to her long-term handling of radioactive substances.

The use of cosmic rays as a high energy particle physics accelerator became an important part of particle physics research over the following decades. Several new particles were discovered using the techniques that Blau pioneered, including the muon and the pion for which more Nobel Prizes were awarded. She was nominated four times for the Nobel Prize in physics and once for the Nobel Prize in chemistry. Cecil Powell was awarded the prize that Blau should have received, not once mentioning her in his acceptance speech. The documents from the Nobel physics committee revealed inaccurate assessments of Blau's research and denied its importance[55]. Sadly, there are no scientific obituaries for Marietta Blau and her contributions to science remain largely forgotten.

Marietta Blau was described as a small, almost fragile woman who was shy and withdrawn in personal dealings, but was very confident as a researcher. In the 1950s, she was also known among her American colleagues for her unconventional driving style: "She stopped when and where it suited her, least of all at red traffic lights and stop signs. Once she came to stand between closed level crossing barriers on the tracks. But the situation ended safely; the train stopped."[56]

---

[55] 'Utredning över Marietta Blaus och Hertha Wambachers till prisbelöning föreslagna arbeten' (Evaluation of work by Marietta Blau and Hertha Wambacher proposed for the prize), A. E. Lindh,1 July 1950, p. 132, Royal Swedish Academy of Sciences, Centre for History of Science, Stockholm

[56] Robert W. Rosner (Hg.): *Marietta Blau – Sterne der Zertrümmerung*. Böhlau Verlag Wien, 2003, page 61.

26. Lise Meitner (1878-1968)

"for the theory and discovery of nuclear fission"

That atoms could fuse together, or gain neutrons to create heavier atoms whilst releasing energy was at the heart of Eddington's, Alpher and Gamow's and Hoyle's most famous research. That atoms could split into smaller pieces also liberating energy was discovered by Lise Meitner, the first female professor of physics in Germany.

Lise Meitner was born into a fairly wealthy Jewish family in Vienna, although her father was a freethinker and religion played no part in her upbringing. With a desire to understand the working of nature she planned on studying at the University of Vienna. Women were admitted to the university from 1897, which was thirty years after women were admitted in the neighbouring Swiss University of Zurich, the first university in central Europe to officially allow women. In order to enrol, Meitner needed to pass her school leaving diploma, the Matura. In 1899 she began private lessons so that she could take the Matura exam. Meitner passed in 1901, cramming eight years of secondary school into two years.

At the University of Vienna Meitner took many lecture courses by the famous physicist Ludwig Boltzmann which she described as *"the most beautiful and stimulating that I have ever heard"*[57]. She carried out a dissertation on heat conduction and became the second woman in Vienna to obtain a doctorate in physics. In 1906 the hot research topic was radioactivity and the structure of matter. Atoms had just been discovered but their structure was still a mystery. Meitner began to carry out research on radioactivity, examining how beams of radiation were deflected by thin foils of metal. She wrote her first scientific publications in 1906 and 1907 on the topic.

---

[57] . Engelbert Broda, „Ludwig Boltzmann: Mensch, Physiker, Philosoph" (Wien: Franz Deuticke, 1955), 9-10.

Options for continuing research in Austria were limited and the centres of research on the new atomic theory of matter were elsewhere in Europe. Funded by her father, in 1907 she moved to Berlin to study under Max Planck. She asked the head of the experimental physics institute if she could participate in doing some research and was introduced to a young chemist Otto Hahn. Hahn was setting up a research laboratory to study radioactivity and needed help from a physicist. The head of the Chemical Institute, Emil Fischer, did not allow women in his laboratory, so they equipped a room in the basement for Meitner which had a separate entrance. She could not even use the restroom in the institute and had to use one at a nearby restaurant[58].

Only after women were admitted to academic studies in Prussia in 1908, was Meitner allowed to work in the research laboratories. In 1912, the Kaiser-Wilhelm Institut für Chemie was opened in Berlin and Meitner continued to work there with Hahn for the next 25 years. Hahn became head of the radiochemistry laboratory and Meitner continued to work without salary. Max Planck recognised her importance and was worried that she might leave so he offered her an assistant position at the Institute for Theoretical Physics marking student papers. Although this was the lowest rung on the academic ladder it was her first paid position and Meitner became the first female scientific assistant in Prussia. The following year she obtained the same associate position as Hahn and the radioactivity research section became the Hahn-Meitner Laboratory. Later it would be divided into two laboratories, each led by one of them.

---

[58] 'Lise Meitner: A Life in Physics', R. L. Sime, 1996, pub. University of California, Berkeley

Hahn was interested in the discovery of new elements, whilst Meitner was more interested in the physics of atomic processes and radiation. Together, they published many research papers until 1914 when war broke out. Hahn was called up for military service and Meitner volunteered as nurse, working with x-rays, for the Austrian army. She still managed to devote some of her time to research and carried out experiments on beta decay and the uranium decay sequence. Immediately after the war she discovered the new element protactinium.

In 1922 Meitner discovered a new process by which atoms emit low energy radiation in the form of a photon and an electron when one of the internal electrons shifts energies to occupy a vacant energy state. The same effect was discovered later by the French physicist Pierre Auger. This effect became known as Auger radiation, but it should be renamed as Meitner radiation. Today, Auger radiation is used as a treatment for cancer. By 1922 Meitner had already published over 40 scientific papers and was promoted to privatdozentin (equivalent to a senior lecturer in the UK, or associate professor in the USA). She gave her inaugural lecture on the topic of cosmic physics, which the newspapers reported as a lecture on 'cosmetic physics'! Meitner's laboratory grew in size and fame over the next decade. Even before she made her most famous discovery in 1938, she had been nominated for a Nobel Prize 13 times yet was never awarded it.

Together with Hahn, she began investigating the radioactive decay of uranium after bombarding the atom with neutrons. Puzzling results emerged but her research was abruptly ended when the Anschluss between Germany and Austria took place in 1938 and she was no longer protected by her Austrian citizenship. It was only a matter of time before she would lose her job and be prevented from leaving Germany. The Dutch scientist Peter Debye arranged for her to flee to the Netherlands and pass Dutch immigration despite having no visa and a worthless Austrian passport[59]. She continued to Denmark where she stayed with Niels Bohr, and a few months later she was given refugee status in Sweden.

In 1938, at the written urging of Meitner, Hahn and his colleague Fritz Strassmann, repeated earlier experiments on the decay of uranium and found that barium was present when uranium was bombarded by neutrons. It was as if the uranium atoms had broken into a much lighter element. Hahn wrote a letter to Meitner to suggest that she *"come up with some sort of fantastic explanation"* since, as he stated, he knew that atoms cannot just split into two. This was indeed astonishing, as all radioactive decay mechanisms known at the time just changed the atomic mass of an element by a small amount by emitting a proton or two. Meitner wrote back to Hahn *"At the moment the assumption of such a thoroughgoing breakup seems very difficult to me, but in nuclear physics we have experienced so many surprises, that one cannot unconditionally say: 'It is impossible.'"*

It was during the winter of 1938 in Sweden, when Lise Meitner and her nephew Otto Frisch, correctly interpreted the uranium experiments and came up with the theory of nuclear fission. Frisch was born in 1904 and was inspired by his aunt to become a nuclear scientist. Frisch recounted the story of how they came up with the discovery during a Christmas walk, with his aunt writing down numbers and equations on scraps of paper they carried. Frisch on skis with Meitner insisting that she could just walk. A decade before, George Gamow had come up with a simple model for the nuclei of

---

[59] 'Lise Meitner, 1878-1968', O.R. Frisch 1970, Biographical Memoirs of Fellows of the Royal Society, vol 16, 404-420

atoms to account for alpha radiation. He likened atomic nuclei to drops of liquid that could be described by a charged fluid with a surface tension. The energy of the alpha particles emitted could be related to the surface tension of the drop. Meitner wondered if perhaps a drop could break into two pieces by first becoming elongated and then constricted in the centre before splitting. She calculated that the charge of the uranium nucleus was almost sufficient to overcome the strength of the surface tension. It resembled a wobbly unstable drop that could be induced to split in two with a small disturbance, such as the impact of a single neutron. She then calculated that after splitting in two, the two separate charged parts would repulse each other at high speed with an energy of around one trillionth of a Joule. Meitner realised the energy would ultimately come from the mass difference between the two fragments. She worked out that the two new nuclei would be lighter by an amount one fifth of the mass of a proton. That doesn't sound like a lot but thanks to Einstein's famous formula, energy equals mass multiplied by the square of the speed of light.

Hahn and Strassmann published their experimental results on uranium decaying to barium without including Meitner. Some scholars have suggested she was excluded because Hahn did not want to risk his laboratory by associating with someone of Jewish descent. Meitner and Frisch wrote a one-page paper to Nature describing their results. The only evidence they had was the barium. That would imply the other part of the nucleus must become krypton, which was soon verified by Frisch. They named the process fission after the same biological process that describes when cells replicate by splitting into two.

It was soon realised that a single kilogram of unstable uranium would release almost 100 trillion Joules of energy if it were induced to decay. That is equivalent to the energy produced by 100 million kilograms of TNT! Scientists realised that if this energy could be harnessed then it could provide a source of energy for mankind. Governments and the military realised that it could be used as a weapon far more powerful than anything previously developed. Meitner declined an invitation to join the Manhattan Project at Los Alamos stating that she would have nothing to do with the construction of a nuclear bomb.

Meitner was horrified to learn of the atomic bombings of Hiroshima and Nagasaki. Ultimately, it was her research that had led to this destructive weapon. Although, like many of the basic discoveries in science, sooner or later someone would have discovered nuclear fission even if Meitner had not. The mechanism behind a nuclear power station or an atomic bomb is simple. Each time a uranium atom splits in two, it converts some mass into energy and releases some neutrons. These neutrons then cause nearby uranium atoms to split into two and a chain reaction occurs. A nuclear power station harnesses the same energy but has to regulate the number of neutrons released to prevent a runaway explosion. Nuclear fission plays an important role in the origin of elements generated in supernova explosions and neutron star mergers. Fission has been speculatively proposed as a possible mechanism by which white dwarf stars can explode and as a means for keeping Earth's interior hot.

Lise Meitner died in her sleep in a Cambridge nursing home aged 89 and was buried in a small village in Hampshire, England. Frisch wrote the inscription on her headstone "Lise Meitner: a physicist who never lost her humanity". Albert

Einstein would often refer to Meitner as the *"German Marie Curie"*. She received several awards for her research, including the Max Plank Medal and the Leibniz Medal. Meitner was nominated a total of 19 times for a Nobel Prize in chemistry and 29 times for a Nobel Prize in physics. In 1944 Otto Hahn was awarded the Nobel Prize for the discovery of nuclear fission, an honour that should have been given to Meitner, or at the very least shared.

27. Yakov Zeldovich (1914-1987)

"for theoretical discoveries in physical cosmology"

The discovery of nuclear fission has affected the lives of us all. It also changed the destiny of numerous scientists in the 1940s who worked on the United States Manhattan project or the Soviet counterpart. Among those was Yakov Zeldovich, the last of the great universalists who made ground breaking discoveries in subjects from physical chemistry, nuclear and particle physics, to gravitation, astrophysics and cosmology. His collected works span many volumes, with over 300 publications in astrophysics and cosmology alone. After Stephen Hawking first met Zeldovich he wrote *"Now I know that you are a real person and not a group of scientists like the Bourbaki"*[60,61].

Yakov Zeldovich was born in Minsk, at the time part of the Russian empire. His father was a lawyer and his mother a translator. Remarkably, he never had a university education[62]. After graduating high school in Leningrad at the age of 15 he joined the Institute for Mechanical Processing of Mineral Resources to train as a laboratory assistant. After an excursion to the Leningrad Physical-Technical Institute his talents were recognised from the questions he asked during the visit. He was immediately invited to work at the Institute of Chemical Physics, initially in his spare time and soon after permanently. He began working on combustion and chemical chain reactions and rapidly gained a reputation as a brilliant young scientist. In 1939 he wrote his doctoral thesis on nitrogen

---

[60] Yu, B. Khariton page 86 in 'Zeldovic: Reminiscences', edited by R.A. Sunyaev, 2004, pub. Taylor & Francis.

[61] Nicolas Bourbaki was the pseudonym of a secretive group of mathematicians who published many books and journal articles.

[62] 'Yakov Borissovich Zel'dovich' V.L. Ginzburg 1988, Biographical Memoirs of Fellows of the Royal Society, Vol. 40, 430-441

oxidation and discovered the chemical pathways by which nitrous oxide is produced during combustion.

That same year, Lise Meitner published her theory of nuclear fission. Zeldovich wrote *"The discovery of uranium fission and the possibility in principle of a chain fission reaction predetermined the fate of the century - and mine as well"*[63]. Soviet scientists began working on nuclear fission, initially as a means for producing energy. Zeldovich developed the theory behind the kinetics of nuclear reactions. He speculated that long ago in Earth's history, a self-sustaining nuclear reaction could have taken place in natural uranium deposits. He realised that today, the concentrations of the isotopes of uranium-238 and uranium-235 have changed so that such a reaction is impossible. Such a natural nuclear fission reactor was discovered in Gabon, Africa in 1972. Deep underground, a billion years ago, a self-sustaining reaction had taken place that lasted for thousands of years.

In 1942, during World War II, Stalin was warned that British and American Scientists had stopped publishing papers on nuclear science. It was obvious that this research had become classified and that they were developing atomic weapons. Stalin immediately ordered all the Soviet physicists to return from military service to work on developing an atomic bomb. Zeldovich played a key role in this project, and he also co-developed the theory behind the hydrogen fusion bomb in which isotopes of hydrogen undergo fusion powered by a nuclear fission bomb. Much of his research from this period of his life remains classified to this day.

---

[63] Selected works of Yakov Borisovich Zeldovich. Volume II. Particles, Nuclei, and the Universe.
Princeton University Press, 1993, page 637.

In 1952 Zeldovich began working on the interactions between particles and the fundamental forces of nature. He proposed laws for the conservation of particle charges and played a key role in the development of the theory of the weak interaction. A decade later, at the age of 50, he switched his focus of research again to astrophysics and cosmology. By this time, he was already a famous Soviet scientist and had been awarded a 'Gold Star of a Hero of Socialistic Labour', the Stalin Prize and a fellowship of the prestigious Soviet Academy of Science. Often, when a scientist completely changes research directions, they fail to have an impact. Zeldovich revolutionised theoretical astrophysics and was one of the key founders of relativistic astrophysics, astroparticle physics and physical cosmology.

It is impossible in a short text to do justice to his volume of research in these areas. In volume II of his selected works, he lists 25 papers on elementary particles and cosmology, 38 papers on general relativity, 30 papers on neutron stars and black holes, 16 papers on the interaction of matter with radiation, 44 papers on the formation of large-scale structure in the universe and 15 papers on observational effects in cosmology. And this is just a selection from many more. I will just mention a few of his major discoveries.

In 1962 he showed theoretically that black holes could form, not only during stellar explosions, but by any mechanism that could compress matter to sufficiently high densities. Zeldovich and his colleague Igor Novikov calculated that under certain conditions, black holes from the size of micrograms to thousands of solar masses could form during the early universe. Such objects are still the subject of research today since they could constitute the missing dark matter. He then began to develop the theory behind how black holes could radiate energy, work that inspired Stephen Hawking's most famous research about which we will learn later.

By the 1950s radio surveys were finding bright sources of emission in all directions. These were given the name quasars but it was not known if these were phenomenon in our own solar system, our galaxy or in the universe beyond. In 1963 the astronomer Maarten Schmidt realised that the light spectrum of quasar 3C 273 implied it was massively redshifted. According to Hubble's law, it was at a distance of over a billion light years away! This implied that it was radiating energy equivalent to that of a thousand galaxies filled with stars. In 1964 Zeldovich suggested that a black hole could be detected by its influence on the surrounding gas which would accrete onto the black hole. He suggested it was the source of energy behind the mysterious quasars. The same idea was developed independently in the same year by the American astrophysicist Edwin Salpeter. It is now the accepted explanation of the power source of quasars – they shine across the electromagnetic spectrum as matter accretes onto the supermassive black holes that lie at the centres of most galaxies.

In 2019 a spectacular astronomical image was made by an international network of radio telescopes which were observing the supermassive black hole at the centre of the galaxy M87. The image is a ring which shows the emission of radio waves surrounding nothing. We think that these photons arise from the glowing accretion disk of matter that surrounds the black hole. This disk forms from material that falls towards the black hole and ends up orbiting it rather than falling directly into it. We can see both the top and bottom of the disk of matter at the same time because the path of the photons curve right around the black hole!

In 1966, Zeldovich and his student Semyon Gershtein showed that if neutrinos have mass they could constitute a large fraction of the matter component of the universe. The idea that dark matter particles could be produced in the big bang predates the time when the evidence for dark matter was accepted. From cosmological considerations they derived an upper limit to the neutrino mass, which at the time was lower than laboratory limits. This was perhaps the first time that it was demonstrated that the universe could be used as a particle physics laboratory. His student went on to show how the number of different types of neutrinos could be constrained by measurements of the helium abundance in the universe. The field of astroparticle physics was born.

Zeldovich regarded one of his main contributions to cosmology was his work on the origin of large structures in the universe. In particular, work on how galaxies form from initially small variations in the matter distribution and how they should be distributed in space. He showed how, as the universe expands, regions with a higher density than average would collapse into large connected structures. In turn, regions of lower density than average would grow into vast empty regions. This supercluster-void network of galaxies is now commonly known as the 'cosmic web', which was discovered with the advent of large galaxy surveys several decades later.

For which discovery should Zeldovich have received the Nobel Prize? Historically, the prize was never given as a 'career award', but for one specific discovery. However, in 2019 cosmologist Jim Peebles was awarded the physics Nobel Prize *"for theoretical discoveries in physical cosmology"* without specifying what those discoveries were. It was clearly an award for his numerous and brilliant contributions to the understanding of the origin of structure in the universe. If Zeldovich had still been alive, I am sure it would have been a joint award which is why I chose the same recognition text as given to Peebles.

Zeldovich learned German and English so that he could read the scientific journals from the West. He despaired at the Western scientists who would not make the effort to read the publications of Soviet scientists, even when translated into English. It was only in the last years of his life that his research achieved grand recognition in the West: he was awarded the Gold Medal of the Royal Astronomical Society, the Dirac Medal and the Gold Medal of the Astronomical Society of the Pacific. Three decades before he had already received the Stalin Prize and the Lenin Prize as well as three Gold Medals of the Hero of Socialist Labour, each with the Order of Lenin. Zeldovich would proudly announce that even Leonid Brezhnev, General Secretary of the Communist Party of the Soviet Union, only had two such medals.

With his vast number of scientific publications and over 20 technical books, you might think that Zeldovich never left his office. That is mainly true, although he did enjoy skiing, swimming and literature. To Zeldovich, an atheist, science was his religion and the most important thing in his life. At the same time, he married three times, following the death of two of his wives, and he had six children, all of whom became physicists. His daughter Marina recalled that he began working at 5am each day and worked seven days a week. And how the 'secretaries' of the KGB used to accompany the family everywhere, even on vacation! Unlike his colleague and friend Andrei Sakharov, Zeldovich never openly challenged the state – perhaps in fear that he would be prevented in carrying out his research. Despite this, Zeldovich was only allowed to travel to the West twice in the last years of his life. His daughter wrote *"...throughout his life he was enchanted with the world, its structure, the beauty of physical theories, seized by the joy of life. ... it was manifest in everything – in his attitude towards science, sports, literature, the theatre, pretty women, and children"* [64].

Zeldovich devoted his life to science and scientific truth and honesty. He always carried a notebook, and would fill it daily with his ideas and thoughts – carefully written, they would later be turned into text books, or pages would be given to one of his many students to follow up on his ideas. He was a small man wearing distinctive thin rimmed round glasses, and with his deep admonishing voice he would be intimidating with his vast knowledge and expectations of others. He used to make bets with his colleagues when they challenged his logic or solutions. A bottle of cognac for the opponent if correct, a bottle of water with a signed label describing the opponent's error if incorrect. His office shelves were filled with labelled bottles of water.

In December 1987 Zeldovich suffered a sudden heart attack and died. The day before he gave a seminar on his latest research and was as energetic as ever[65]. That same year, aged 73, he had already published over a dozen scientific papers. Amongst his last, published after his death were two studies entitled *"The spontaneous creation of the universe"* and *"Is the formation of the universe from nothing possible?"* I wonder what

---

[64] M. Ya. Ovchinnikova pages 61-68 in 'Zeldovic: Reminischences', edited by R.A. Sunyaev, 2004, pub. Taylor & Francis.

[65] 'Ya B. Zeldovich', V.I. Goldanskii, 1988, Obituary in Physics Today, 41, 12, 98

contributions to science he would have made if he had lived to be a hundred.

28. Fabiola Gianotti 1960- and 3000 others

"for recreating the conditions within the first second of the big bang and for discovering the Higgs boson"

The microwave background photons give us a picture of what the universe looked like 380,000 years after the big bang began. It also revealed the physical conditions of the universe at that time. The origin and abundance of hydrogen and helium gives us information about the first few minutes. Probing even earlier epochs had to wait until the construction of giant particle accelerators, and the greatest of those is the Large Hadron Collider (LHC) at CERN. In 2009, Italian physicist Fabiola Gianotti took charge of its largest team, a 3000-member research group that constructed the ATLAS detector which detected the mysterious Higgs boson particle in 2012. In 2015 she became the director general of CERN.

As a child, Fabiola Gianotti dreamed of becoming a ballerina, and then trained to become a classical pianist. But she was also inspired by science after reading a biography of Marie Curie, and had to decide – after all, it is hard to do more than one thing exceptionally well in life, so she chose to keep music and dance as a hobby and earned a PhD in Physics in 1989 from the University of Milan. She joined CERN as a research physicist in 1994 a year before the budget for the LHC was approved. The motivation for the LHC was to probe the interaction between particles with energies comparable to those during the earliest moments of the big bang.

Using the equations of Friedmann and Lemaître (which tell us how the size of the universe changes with time), we can calculate physical conditions such as density or temperature at any epoch in the history of the universe. As the amount of space between particles of matter increases, their relative speeds decrease and fewer collisions take place between the particles. Consequently, the temperature and density of the universe both decrease with time. Photons also lose energy as the universe expands because their wavelengths stretch along with the space and their energy decreases. It is as if photons are somehow pinned to space and as space expands, the photons transfer their energy into the newly created space.

We can also extrapolate backwards billions of years to a time that is just a tiny fraction of a second after the big bang has started. The average temperature of the universe today is about 2.7 Kelvin (−270.5 degrees centigrade). That is the temperature you would find if you put a thermometer in empty space well away from a nearby star. It is the temperature of the cosmic microwave background photons today. If you could put the same thermometer at the same point in space in the early universe, say one second after the big bang, you would measure a temperature of 10 billion Kelvin! But as we get closer and closer to 'the beginning' our understanding becomes less and less certain. The reason for the lack of a complete description of the big bang is that the conditions become so extreme that we start to find major gaps in our knowledge of fundamental physics. We simply do not have a unified theory of the four forces that we know occur in nature, and we do not understand how quantum mechanics and gravity work together under the most extreme conditions.

A millionth of a second after the big bang the universe was filled with fundamental particles, such as quarks, electrons, neutrinos and their antiparticles. The temperature of the universe was still too high to allow the quarks to bind together to form protons or neutrons, and space was filled with a hot quark-gluon plasma. This is the era of time that the LHC was designed to probe, by colliding particles with energies which are similar to the interactions that occurred during this epoch.

The LHC is one of the most sophisticated scientific experiments ever undertaken by humankind. It took over a decade to construct using the combined efforts of tens of thousands of scientists and engineers from over a hundred countries around the world. The experiment consists of a circular 27-kilometre-long tunnel passing under the Swiss and French countryside, lined with superconducting magnets which accelerate the protons to a speed that is just 3 metres per second less than the speed of light (almost 300 million metres per second).

The tunnel itself is about as wide as a train tunnel, but the particles are actually accelerated by an electrical field in a tube no wider than a fire hose. The beam tube holds an ultrahigh vacuum, as empty as interplanetary space; otherwise, the protons would collide with molecules of air. The proton beam itself consists of about three thousand packets each containing a hundred billion protons: Half of the packets are sent clockwise and the other half counter-clockwise. The actual clouds of particles are about a millimetre across and weigh less than a single blood cell, but at full speed they have the energy of a four-hundred-ton train travelling at top speed! Remember, energy is mass. To keep the protons within the ring, there are over nine thousand magnets along the route which focus the protons and eventually concentrate the packets to the width of a human hair before they collide. The superconducting magnets each weigh over 27 tons and are cooled down to –271.3 degrees centigrade to increase their efficiency. The magnets need to be extremely powerful since, as the protons move faster, their inertial mass increases 10,000-fold. That's a consequence of Einstein's general relativity. Speed, mass, energy and time are interdependent: the faster you travel, the more you weigh and the more your clock slows down.

It would take the Space Shuttle about five seconds to travel a distance the length of the LHC tunnel – it travels 10 times faster than a bullet. That is impressive, but after being accelerated for about an hour, the packets of protons reach their top speed: they are travelling so fast that they travel around the 27-kilometre tunnel over 11,000 times in a single second!

Finally, when they cannot be made to move any faster, the two particle beams are made to crash into each other inside one of the giant detectors which occupy an underground cavern as big as a cathedral. When two packets of protons collide, there will only be about 20 collisions between a trillion crossing particles. But since the packets cross 11,000 times per second, hundreds of millions of collisions take place. These collisions between particles occur at the highest energies ever achieved on Earth, and are comparable to those taking place during the first millionth of a second of the universe. The giant ATLAS detector records all of the details of the collisions, generating 15 million gigabytes of information per year.

The ATLAS detector is the size of a seven-story building. It contains concentric layers of instruments that records the trajectory, momentum and energies of particles produced in the collisions. Over a billion particle interactions take place in the ATLAS detector every second, a data rate equivalent to 20 simultaneous telephone conversations held by every person on the Earth. Only one in a million collisions are flagged as potentially interesting and recorded for further study. The detector tracks and identifies the particles that are created in the collisions to investigate a wide range of physics, from the study of the Higgs boson and top quark to the search for extra dimensions and particles that could make up dark matter. A second experiment, the Compact Muon Solenoid (CMS), has a similar design and can operate at the same time as the ATLAS experiment

One of the main goals of the LHC was to detect something called the Higgs boson. Bosons are subatomic particles that are responsible for transmitting the fundamental forces of nature – the electromagnetic, weak and strong forces. They act like messenger particles when normal atomic matter interacts. There are five basic bosonic particles: photons, gluons and the W, Z and Higgs bosons. All of these except for the Higgs boson had previously been detected in experiments. Detecting it completed what is called the standard model of particle physics. From the billions of collisions that occur, the signature of a Higgs boson was expected to be seen only once every few hours. By 2012 enough data had been collected in the ATLAS and CMS experiments to determine that the Higgs particle existed and the results were announced to the world by Gianotti.

Gianotti is co-author on over 500 refereed publications, and many of those, such as the Higgs discovery paper, have a list of several thousand co-authors. And herein lies a problem with today's experiments which rely on thousands of technicians, engineers and scientists. No longer can a small team of researchers make a new discovery – most of the easier discoveries have already been made. Gianotti has received many honorary degrees and has been awarded several prizes, such as the Enrico Fermi prize of the Italian Physical Society and the Special Breakthrough Prize in Fundamental Physics. It seemed all set that Gianotti and her team, and that of the CMS experiment, should be recognised for the discovery of the Higgs boson. In 2013, following the ATLAS discovery, the Nobel Prize in Physics was awarded. But not to any of the scientists involved with the actual discovery - the prize was shared between the physicists François Englert and Peter Higgs for their theoretical prediction of the Higgs particle. Even this award was controversial since the theory behind the Higgs particle was developed by at least six different physicists.

Whether or not Gianotti should be awarded a Nobel Prize is a matter of debate since with the current rules of the Nobel Prize, the team behind the experiment would not be recognised. But that has not prevented such awards in the past. In 1984 the Nobel Prize was given to two scientists, Carlo Rubbia and Simon van der Meer, for their role in the discovery of two other particles at CERN, the W and Z bosons. These discoveries also involved teams of hundreds of researchers. When asked whether she was disappointed in not sharing the Nobel Prize Gianotti stated *"Certainly not for me and I think many of my colleagues shared the same feeling. When you discover something, first of all you are very much rewarded by the discovery itself."*[66] Indeed, awarding the Nobel Prize in its current format makes little sense for such discoveries. It is often the case that no one scientist, or three scientists, can be distinguished – such research is a collective activity.

Could we go further back, would it be possible to construct an even larger LHC to probe the first instant of time in our universe? The next instant of time where 'new physics' happens is predicted to have occurred around 10^{-32} seconds after the big bang. It is at this time that it has been proposed that an instability in the vacuum energy drove the rapid expansion of the universe. We will learn later about the researchers who came up with this theory of the very early universe. The problem now is that in order to test earlier

---

[66] https://www.jotdown.es/2017/12/fabiola-gianotti-eng

phases of our universe we would need to reach energies far higher than particle colliders on Earth can ever achieve with current technology. Gianotti, as head of CERN, is proposing a Future Circular Collider which would cost well over 20 billion Swiss Francs. However, there is little scientific justification beyond the hope of finding something unexpected. The next generation particle colliders could increase the energy of the particle collisions by only a factor of about thirty. That is still probing the regime where the universe was in its quark-gluon plasma stage. We would have to build particle accelerators the size of the solar system to reach the energy scales where we might expect new phenomenon.

29. Jocelyn Bell Burnell (1943-)

"for the discovery of neutron stars"

A white dwarf star is not the ultimate dense state of matter, far from it. The idea that stars might end their lives by collapsing into a new state of matter, called a neutron star was first proposed by Walter Baade and Fritz Zwicky in 1934. Neutron stars were discovered in 1967 by the British astronomer Jocelyn Bell. These objects contain more than the entire mass of our Sun compressed into a sphere just 20 kilometres across. A thimbleful would have a mass of a billion tonnes. That sounds astounding and it is.

Jocelyn Bell was born in Northern Ireland in 1943. When she was aged 11, at school the girls were put in 'domestic science class' where they were supposed to learn cookery and needlework, whereas the boys were put in the science lab to learn physics and astronomy. Twenty minutes into the class Bell said to the teacher *"I think I'm in the wrong place"* and asked to be moved to the science class[67]. By the end of the year, she came top of the class in the exams achieving a score of 97%. She only got one question wrong, which was to state the speed of light. She wrote down the correct answer she had memorised, 186,000 miles per second, but then she thought that's really fast, crossed out seconds and wrote hours. Her father had taught her to question everything and to see if her answers made sense.

Later, at high school in York aged 15 her father bought her an astronomy book by Fred Hoyle and she decided she wanted to become an astronomer. Her hopes were partially dashed when she visited the Armagh observatory, which was designed by her father who was an architect. One of the astronomers told her that if she wanted to become an astronomer she would have to be good at staying up all night.

---

[67] American Institute of Physics, Oral History Interviews, by D. DeVorkin 2000.

Bell became rather depressed since as a teenager she recalls loving her bed and sleep. Instead, she decided she would become a 'radio astronomer' since you could observe the cosmos during the daytime. She sent a letter to the famous astronomer, Bernard Lovell at Jodrell Bank, asking him what she needed to study at university to become a radio astronomer, and Lovell wrote back telling her physics.

At university in Glasgow she was the only female in the honour's physics class of fifty students – she was often advised to not bother studying physics and recalls the tradition that whenever a woman walked into a lecture hall all the guys would stamp their feet and whistle. She again excelled at the exams, often coming top. She succeeded in obtaining a PhD position with the Cambridge radio astronomy group in 1965, led by Martin Ryle. Her supervisor was Antony Hewish who was building new instrumentation to observe quasars – Bell helped in the construction, driving stakes into the ground in a field over twice the size of a football pitch, connecting long series of wires that would be the antenna. After two years of construction, it worked the first time the radio telescope was switched on!

As the Earth spun, the radio telescope measured radio signals from whatever sources appeared in the sky overhead. An automated pen would record any signals on a long paper chart – each day would produce about 30 metres of paper measurements. Bell operated the equipment and analysed the chart data each night. With a radio telescope you have to be careful of man-made interference. For an entire week the signals were swamped by the police who were using the same radio frequencies as Bell could observe – until she managed to persuade them to use a different frequency. Then, one night she noticed an unusual source of radio emission that seemed to be switching on and off with a period of about one second. She noticed it again and again on following nights.

Bell reported her findings to Hewish who said that this must be some human made signal since an astrophysical source pulsing with a frequency of one second would have to be smaller than one light second across – far smaller than even a star. Bell realised that the object always appeared in the sky at the same sidereal time, which measures when objects appear in the same place in the sky, so it must be from a source beyond our Earth. After looking through about five kilometres of pen-chart data she found another strange pulsing object in the sky. The pulsating radio source discovered by Bell was later called a pulsar by a journalist. The name stuck. Bell had discovered a neutron star that was emitting pulses of radio waves because of its rotation.

For stars with an initial mass between 8 and 25 times that of the Sun, their central regions are hotter and denser. When their fusion ends and gravitational collapse ensues, gravity is strong enough to squeeze atoms even closer together, so close that the nuclei are effectively touching. The end state is a neutron star which has the density of atomic nuclei and further collapse is halted thanks to a new form of quantum mechanical pressure provided by the neutrons. They are thought to have a crust of atomic nuclei, but inside the neutron star under these extreme conditions, the protons have all been transformed into neutrons via the capture of electrons. A teaspoon of a neutron star on Earth would weigh as much as 300 million tonnes. That is equivalent to squeezing the entire human population into a space the size of a single sugar cube!

Neutron stars rotate very fast: up to several hundred times a second. This is a consequence of the conservation of rotational energy – as a rotating star collapses to a smaller state, its rate of rotation must increase, just like a spinning ice-skater that spins faster when they move their arms inwards. From looking at how the sunspots move, Galileo realised that the Sun rotates about once per month. If it were to collapse to the size of a neutron star it would spin a billion times faster and neutron stars have actually been observed spinning this fast. The rate at which they spin can be measured by studying pulsars. A pulsar is a neutron star which emits regular pulses of radiation in our direction.

In the same way that the neutron stars spin faster as they collapse, their magnetic fields also become stronger. The charged particles on the surface of a neutron star interact with its magnetic field. As a result, it gives rise to electromagnetic radiation – a beam of light from its north and south poles is emitted into our galaxy. Some pulsars wobble as they spin and we only observe the light when it is pointing towards us, rather like the beacon of a lighthouse. The time between the pulses of light are measured to be as regular as our atomic clocks. Bell had observed the pulses of light in the radio frequencies from such a rapidly spinning neutron star.

Bell recalls entering Ryle's office and finding a meeting taking place about how to make the discovery announcement to which she had not been invited. Her results were sent to the journal Nature with five authors. She was listed second, her PhD advisor as first author and the boss of the institute Martin Ryle was also included. In 1974 the Nobel Prize in physics was awarded to Hewish and Ryle for their work in radio astronomy and the discovery of pulsars. Bell was excluded, much to the wrath of Fred Hoyle. Hoyle accused Hewish of stealing Bell's data, and was extremely critical of the Nobel Prize committee. Because of British libel laws Hoyle had to publicly apologise to Hewish, and his criticism of the Nobel committee may have contributed to the fact that Hoyle himself never received the Nobel Prize. However, Hoyle was correct in his criticism – not awarding Jocelyn Bell the Nobel Prize is certainly one of the most shocking oversights noted in this book. Bell constructed the radio telescope, took the observations, made the discovery, fought with her senior colleagues about its reality and confirmed its presence and nature as a new type of extreme astrophysical object. In 1977 she commented on her lack of recognition *"...I am not upset about it – after all, I am in good company, am I not!"*[68]

Following her discovery, she taught at the University of Southampton before becoming a professor at University College London. She subsequently received many other prizes and awards for her research, including the 2018 three-million-dollar Special Breakthrough Prize in Fundamental Physics. Bell donated all the money to fund women and under-represented ethnic minority and refugee students to become physics researchers. In a BBC interview Bell stated *"I found pulsars because I was a minority person and feeling a bit overawed at Cambridge. I was both female but also from the north-west of the country and I think everybody else around me was southern English. So I have this hunch that minority folk bring a fresh angle on things and that is often a very productive thing. In general, a lot of breakthroughs come from left field."*[69]

---

[68] 'Petit Four', after dinner speech by J. B. Burnell 1977, Annals of the New York Academy of Sciences, vol 302, 685-689

[69] Pallab Ghosh: Bell Burnell: Physics star gives away 2.3million pound prize, BBC News, Sept 6, 2018

30. Shiv Kumar (1939-)

"for predicting the minimum mass of a star, the existence of brown dwarfs and population III stars"

A star is a giant ball of mainly hydrogen and helium plasma which undergoes sustained nuclear fusion. But many astronomers prefer to name things according to how they form rather than the physical processes they undergo. That's why we have objects such as white dwarfs that are frequently called white dwarf stars, even though they are not undergoing nuclear fusion. In 1963 the Indian astronomer Shiv Kumar speculated that the same process of gravitational contraction that creates stars from vast clouds of gas and dust would also frequently produce smaller objects. He calculated that below a certain mass, a star would form but never shine because it could not undergo nuclear fusion.

Shriv Kumar was born in 1939 in Bannu, India – a region that is now part of Pakistan. After studying mathematics at Lucknow University in India, he joined the University of Michigan to carry out his PhD on the atmospheres of stars. He then spent most of his career at the University of Virginia. In the 1940s the American astronomer Henry Russell had speculated that stars might exist that were too small to ignite nuclear fusion. There was little other mention of such a possibility until Kumar began to perform calculations of the evolution of stars of lower and lower masses. Starting in 1958, he kept lowering the mass of his models, until by 1962 he discovered that below a certain mass, fusion would not occur. For fusion of hydrogen to begin, temperatures of several million degrees are needed. That smallest mass that a star can undergo hydrogen fusion is about 80 times the mass of Jupiter, or about eight percent of the mass of our Sun.

Kumar published his results on the minimum mass of a star in 1963. He then calculated how a 'failed star' evolves over time. It begins its life hot because of its gravitational collapse.

As it cools, it becomes denser, eventually reaches a size that is similar to that of the planet Jupiter, completely black as it emits no more radiation. Later calculations showed that the most massive of Kumar's failed stars could attain temperatures at their core to fuse together deuterium, an isotope of hydrogen. This requires slightly lower temperatures which is why deuterium is the fuel of choice in experimental fusion reactors. However, deuterium is rare and a small fraction of the matter inside stars, so this fuel source could only last for a short period of time.

Why do we call such objects brown dwarfs? Brown is not a colour of the rainbow which arises from specific wavelengths of light. The colour brown is made by combining red, blue and yellow. Kumar originally called his failed stars black dwarfs, so how did the modern name of brown come about? The problem was that the name black dwarf was already in use in the 1960s. It was the name given to the white dwarf remnant of a star, that cools down over cosmic time and would eventually appear black. This process takes longer than the age of the universe, so no black dwarfs exist yet. Because Kumar's objects should glow in the infra-red wavelengths, red dwarf would have been a better term, but the name red dwarf was already used to describe low mass stars. In 1975 the astrophysicist Jill Tarter suggested calling Kumar's objects brown dwarfs and the name stuck.

All objects that form from the gravitational collapse of matter, from planets to stars, will end up hot due to the conversion of gravitational potential energy into kinetic energy. In a planet, this heat energy slowly radiates away over time. In a star, a new form of energy kicks in – nuclear fusion. A brown dwarf star is more like a planet, starting out hot and radiating heat energy in the infra-red wavelengths. As it cools down the light becomes fainter and moves to even longer wavelengths. It was realised they would be hard to detect and despite many observational searches between the 1970s and 1990s, no brown dwarf was discovered. For this reason, it was often speculated that such objects could make up the missing mass in our galaxy.

The crucial test of whether a faint object is a brown dwarf and not an extremely faint red dwarf star, is the presence of the element lithium created during the big bang. A star undergoing fusion will deplete its primordial lithium within about 100 million years. In brown dwarfs, lithium will persist. Another test is the presence of methane in their atmospheres. This is mostly destroyed in the hot atmospheres of stars. The hunt was on, for objects far fainter than the Sun with lithium or methane in their atmospheres.

In 1995 a conference in Florence, Italy, took place on the topic of stars where two independent groups of astronomers announced their discoveries of brown dwarfs. At the same meeting, Michel Mayor announced his results on the discovery of an extrasolar planet orbiting another star. That must have been an exciting conference! One of the newly discovered brown dwarfs was Gl 229B, an extremely faint companion to the red dwarf star Gl 229. The search for brown dwarfs nearby known stars was motivated by the fact that most stars form together with a companion star. Gl 229B emits one millionth of the luminosity of our Sun, has a surface temperature of 1000 Kelvin (the left-over heat from its formation several billion years ago), and its spectrum reveals the presence of methane.

In 1964 Kumar pointed out that his brown dwarfs could make up a significant amount of the mass in our galaxy – perhaps far more than the visible stars. Since 1995 searches for brown dwarfs have progressed and several thousand brown dwarfs have been discovered. Because they are so faint it is difficult to estimate their numbers, but between 25 and 100 billion such objects are thought to exist in our galaxy alone! But because of their low mass, this has ruled out the idea that they could make up the missing matter.

Observations of known brown dwarf candidates have revealed varying infrared emissions that suggests relatively cool, opaque cloud patterns obscuring a hot interior that is stirred by extreme winds. The weather on such bodies is thought to be extremely strong, far exceeding Jupiter's famous storms. There are ongoing searches for planets orbiting such objects – such worlds would never receive the light that Earth receives and are therefore most likely devoid of life. But, in the far future, long after all the stars in our galaxy stop shining because they run out of nuclear fuel, rare collisions between brown dwarfs may take place. Roughly every 10^{18} years, such a collision could result in a merger between such objects, creating a remnant that has sufficient mass and temperatures to undergo fusion – a red dwarf star. Such a star would shine alone in our galaxy for a trillion years, before another lengthy period of darkness ensued.

What is the difference between a brown dwarf and a gas giant planet such as Jupiter? Not much, only the method of formation is different. It is not known how small a brown dwarf could be. Currently, the International Astronomical Union considers an object above 13 times the mass of Jupiter (the limiting mass for thermonuclear fusion of deuterium) to be a brown dwarf, whereas an object under that mass (and orbiting a star or stellar remnant) is considered a planet. The minimum mass required to trigger sustained hydrogen-burning, about 80 times the mass of Jupiter, forms the upper limit of the definition.

In a series of papers from 1967 to 1974, Kumar wrote about the origin of Jupiter. He speculated that Jupiter began as a small rocky object within the protoplanetary disk that grew to its current size by accreting dust and gas. Today, the so called 'core accretion' model is the most favoured explanation as to how gas giant planets such as Jupiter and Saturn form. Kumar's initial works on this seem to have been long forgotten, as was his suggestion of the existence of a third population of stars.

In the early 1960s Allan Sandage published his measurements of the age of the galaxy. In a paper that most astronomers seem to have completely forgotten since it does not have a single citation, Shiv Kumar in 1962 published 'On the age of the galaxy'. In this he points out that the age Sandage derives is a lower limit to the age of the galaxy. At the time, Walter Baade had divided up the stars into so called young population I and old population II stars. Kumar argues that the population II stars observed by Sandage have elements heavier than helium in their spectra. He correctly speculates that these elements must have been synthesised in an even older generation of stars he calls 'first generation stars' that must have formed from pure hydrogen and helium – since that's all that was made in the big bang. These stars would have then exploded as supernova, enriching the galaxy with newly synthesised elements. He calculated that the true age of our galaxy could be six billion years older than Sandage measured. Today we call such stars, population III stars and they are a hot topic of research. Calculations reveal that they should live short lives, we do not expect to find any such primaeval objects in the galaxy today. But they play a crucial role in the history of galaxy formation. They could be detected, but it would take an even larger telescope than the James Webb Space Telescope.

Despite never receiving any awards or prizes for his work, and the fact that some of his ideas are long forgotten, at least Kumar's research on the minimum mass of stars and brown dwarfs has been recognised. In 2012 an international conference took place organised by the Max Planck Institute for Astronomy in Heidelberg titled '50 years of brown dwarfs', at which Shiv Kumar was guest of honour and spoke about his early work.

31. Aleksander Wolszczan (1946-) and Dale Frail (1961-)

"for the discovery of extra-solar planets"

The discovery that nearly every star we can see in the night sky is orbited by its own worlds, extra-solar planets, is certainly one of the most stunning of all astronomical discoveries. The 2019 Nobel Prize in physics was awarded to Swiss astronomers Michel Mayor and Didier Queloz *"For the discovery of an exoplanet orbiting a solar-type star"*. It was well deserved, but it was not the first discovery that extra-solar planets exist. That was made three years earlier than Mayor and Queloz, by the Polish astronomer Aleksander Wolszczan and his colleague American astronomer Dale Frail. The only difference was the planets they discovered were not orbiting a star, but the remnants of a dead star – a pulsar.

Aleksander Wolszczan was born in Poland and studied radio astronomy at the Nicolaus Copernicus University in Toruń. He worked at Cornel University and Princeton, before becoming a professor at Pennsylvania State University. In an online interview for that university, he stated

"I think I have always been fascinated with time and space and how that works and what is our relationship to time and space. That's why I like mountains. That's why I like to be on the shore of a big ocean and just watch it. I don't have to do anything. It's simply fascinating and kind of mind gripping. And, the same is true for the sky. It's bottomless."

Wolszczan's first scientific publication in 1974 reports on observations of pulsars. It was a topic on which he spent his entire career. This represents a more typical career of a scientist – they usually specialise in one or perhaps two topics. The likes of Silk and Zeldovich are the exception, especially amongst today's researchers. But it was this specialisation, knowing everything there is to know about a specific research area, that enabled his famous discovery.

We already heard about the discovery of pulsars by Jocelyn Bell. When a massive star runs out of fuel and stops shining there is nothing to prevent gravity from pulling all the material towards its centre. The core of the star collapses in seconds and the resulting shockwave causes a spectacular explosion and the outer layers of the star are literally blasted away into space – a supernova. The collapsing core can form a neutron star, or even a black hole if the initial star was sufficiently massive. A spinning neutron star can emit regular pulses of radio waves which are generated and focussed by its intense magnetic field. The narrow beam of radio waves sweeps the sky like a lighthouse, but much faster – from seconds to milliseconds because that's how fast these remnants of dead stars spin. If we are in the path of the beam we can observe the pulses of radio waves as the radio beam spins in and out of view

The rotation of the neutron star, or pulsar, is so regular that the interval between pulses is as regular as an atomic clock. Anything orbiting the pulsar causes the pulsar to wobble backwards and forwards, slightly changing its distance from Earth. This changes the arrival time of the pulses which can be measured to a very high accuracy. The position of the pulsar can then be determined to an astonishing precision of 300 metres which is accurate enough to detect an orbiting planet.

This relies on the same physics as the basis of Mayor and Queloz's detection of the first planet orbiting a star. Because a planet has mass it also affects the object it orbits. Take for example our Sun. The gravitational attraction of the planets also pulls on the Sun, causing it to wobble in space. The Sun and planets actually orbit the centre of mass, the barycentre of the solar system, which varies from between one and two solar radii from the centre of the Sun. Because Jupiter is so massive it is responsible for most of the Sun's wobble. But all the planets have some effect on the Sun causing it to spiral around in a complex but predictable fashion. These wobbling motions of a distant star with planets can be detected in two different ways, by measuring its change in position or by measuring its changing speed.

Since the resolution of telescopes is too low to see the change in position of a star, astronomers can measure its change in speed to an accuracy of metres per second, sufficient to detect the effects of an Earth-like exoplanet. But the position of a pulsar can be determined so accurately that its variation in distance reveals the presence of any orbiting worlds. Wolszczan first used the famous Arecibo radio telescope to observe the pulsar *PSR 1257+12* which lies 1000 light years away. He noticed strange irregularities in the arrival times of the radio pulses, but Arecibo could not pinpoint the exact location of the pulsar. He needed this information to be able build a model in which orbiting planets could explain the data. He contacted Dale Frail, an astronomer at the National Radio Astronomy Observatory. Its facilities consist of 27 radio dishes that work together and can provide data to a higher precision. Frail faxed the results to Arecibo with a note *"Don't find any planets!"*. That's because the preceding years has seen several famous announcements of planet discoveries that were later retracted! Frail soon got a message back *"Two planets"*!!

After Wolszczan first looked at the data he later said *"I basically walked out of the office. I needed to calm down a little bit. Up to that moment so many planets had been detected and then retracted that you would sort of approach someone and say, 'Hey, I found some planets.' When somebody said something like that to me, I would say, 'Okay, show me your data'".*[70]

In 1994 Wolszczan was able to confirm his results by looking at the gravitational effects that the planets had on each other. The two planets were not much larger than the Earth, but his analysis also revealed the presence of a third planet, not much larger than the mass of our Moon. The pulsar timing technique is so powerful that the Moon sized exoplanet that they discovered is still the smallest known exoplanet.

It is a puzzle how those planets got there. When a massive star explodes and ejects its outer layers, over half of its mass is blown away into space and any orbiting planets suddenly feel a weaker gravitational pull. The neutron star remnant that is left is not sufficiently massive to keep them in orbit and the planets should be lost into galactic space. Perhaps these planets formed from the supernova debris - a cosmic birth of new structures from old. If there were life on a planet it would not survive such a nearby supernova explosion, but it is unlikely that they ever hosted life. Massive stars use up their fuel much more quickly and shine for less than 30 million years. This is insufficient time for a planet to become habitable and for life to evolve. The pulsar doesn't provide energy that could facilitate simple life. The surface temperatures of these planets are extremely cold and no known liquids could exist there.

Wolszczan and Frail published their results in the journal Nature in 1992, in a paper titled 'A planetary system around the millisecond pulsar PSR1257 + 12'. In a strange twist to this story, they submitted their results to Nature in 1991 and Wolszczan was due to give a talk at the January 1992 meeting

---

[70] 'Planets from the very start', 1997 interview of Alexander Wolszczan by Charles DuBois for Penn State University.

of the American Astronomical Society. This is a large prestigious conference attended by hundreds of astronomers and the press. Just before his presentation, the radio astronomer Andrew Lyne was scheduled to talk about a detection of planets orbiting pulsars which he had published in 1991. Wolszczan thought he would just be the second scientist to announce such a discovery. However, to the astonishment of the audience Lyne explained during his talk that he had made an error in his calculations and retracted his results. Lyne had not measured the position of the pulsar accurately enough. But his frank admission was acknowledged as a demonstration of scientific integrity and the audience stood up to give him a standing ovation. Next to speak was Wolszczan, who was also shocked by the announcement, but even more so that he was to present his new results to a now sceptical audience!

In yet another twist to this story, in 1988 three Canadian astronomers led by Gordon Walker and Bruce Campbell published the results of a study in which they had searched for exoplanets by measuring the wobble of stars. They had a hint of a positive detection of a planet orbiting the star Gamma Cephei. But they basically excluded the possibility in a later paper published in 1992, putting the results down to the rotation and oscillations of the star. It was only in 2003, that another team of astronomers confirmed that indeed, Gamma Cephei does indeed have a planet a few times the mass of Jupiter. And in 1989, another team led by David Latham of the Harvard-Smithsonian Center for Astrophysics reported evidence for a planet 13 times the mass of Jupiter orbiting a star, but noted that it was probably a brown dwarf star. Indeed, at that mass we have heard that a brown dwarf can undergo deuterium fusion.

The discovery of exoplanets has led to a new field of astronomy – the characterisation of planets orbiting other stars and the search for life that they may host. One of the main scientific goals of the next generation telescopes, such as JWST and ESO's 39 metre telescope, is to conduct these searches.

Sometimes I do not envy the Nobel Prize selection committee. But only three names could be selected and at least four people deserved the Nobel Prize. Since the pulsar is essentially a dead star, and the Gamma Cephei observations were insecure, they awarded the Nobel Prize to Mayor and Queloz whose data were quickly confirmed and clearly showed the existence of a planet orbiting another star. Perhaps Wolszczan should have been awarded the prize without Frail. After all, it was he who first found irregularities in the data and performed the modelling calculations that showed the existence of planets. But without knowing the accurate position of the pulsar that Frail provided, his results would have been worthless. And that is how science works.

32. Carl Sagan (1934-1996)

"for determining the past and future of the Earth in the face of an evolving Sun, and for inspiring us all"

Carl Sagan is well known for the television series Cosmos, the most widely watched series in the history of American public television and which had over half a billion viewers worldwide. The accompanying book was on The New York Times bestseller list for 70 weeks and at the time was the best-selling science book ever published. He wrote many other popular science books, and if I could recommend one it would be 'The demon haunted world: science as a candle in the dark' in which he explains the scientific method, critical thinking and how to detect pseudoscience. In the book Sagan argues quite beautifully that *"Science is not only compatible with spirituality; it is a profound source of spirituality"*[71]. His 1985 science fiction novel 'Contact' was turned into the movie with the same name. He was without question the greatest populariser of science the world has seen. However, even we scientists tend to forget his great contributions to our understanding of the solar system. Sagan published over three hundred peer reviewed scientific papers and in one of those he calculated the past and future of our Earth and Sun.

Sagan was born in a working-class neighbourhood of New York and by the age of 12 he expressed his desire to become an astronomer. After excelling at school, he obtained a scholarship to study at the university of Chicago and stayed there to carry out his PhD research in astronomy. It was there that he met his first wife, Lynn Margulis, who described their early relationship: *"He was tall, handsome, exceedingly articulate...His gift of gab fascinated me ...His love for science was*

---

[71] The demon haunted world: science as a candle in the dark, 1995, Carl Sagan & Ann Druyan, pub. Random House, Page 29

contagious." [72] Margulis would go on to become a famous evolutionary biologist. If I were to write about unrewarded biologists who deserved a Nobel Prize it would include Margulis. She proposed the idea that multicellular life with its unique compartmentalised cells arose thanks to a symbiosis of microbial cells – an evolutionary step that was one of the most dramatic events in the history of life on Earth.

Later Sagan would move to Harvard University where he spent five years as an untenured professor. Surprisingly, despite his already outstanding contributions to research, Harvard did not promote him to full professor, partly thanks to a negative reference letter from the Nobel Prize winning chemist Harold Urey. Later in his career, Sagan would also be denied membership to the National Academy of Sciences – the nomination needed two thirds of the members to vote him in but fewer than half did. These snubs are widely attributed to the fact that many scientists are jealous of those who write books and popularise science and they have an unwarranted low opinion of their scientific achievements. This phenomenon is now known as 'the Sagan effect'. From my own experience I can certainly attest to the petty jealousy of some scientists – my advice is to wait until you have tenure before writing a book or regularly appearing on television! Sagan moved to Cornell University where he became director of the Laboratory for Planetary Studies. At Cornell his research achievements and outreach activities were both appreciated and he spent the rest of his career happily there.

For his PhD thesis, Sagan predicted that the temperatures on Venus could be extremely high thanks to a runaway greenhouse effect. He would later use radar observations of

---

[72] Symbiotic Planet: A New Look at Evolution, 1999, Lynn Margulis, pub. Basic Books, page 17.

Venus to argue that it had a surface temperature of over 400 degrees centigrade. This was confirmed by the Mariner 2 space probe which carried out the first successful close-up observations of another planet when it flew past Venus in 1962. He used Venus as an example to warn of the dangers of what could happen on Earth if global warming became out of control.

Sagan pioneered the field of exobiology and the search for life on other worlds. He carried out experiments to show that the building blocks of life, such as amino acids and nucleotides, could be synthesised by exposing chemicals to ultraviolet radiation. He also wrote about the possibility that life could exist in the thick atmosphere of Venus where 50 kilometres above the ground the temperature and pressure match that of Earth's atmosphere. The idea was mostly ignored by the scientific community. However, in 2021 astronomers discovered the molecule phosphine in the atmosphere of Venus. Its abundance is too high to be explained by natural causes, and it is found only in Earth's atmosphere because it is produced by life. This has renewed interest in Venus and a number of new exploratory space missions are being planned to study its atmosphere and surface.

Sagan was an advisor to NASA from its inception in the 1950s. He played a leading role in the Mariner, Viking, Voyager and Galileo missions to other planets. One of the most famous photographs ever taken is an image of the Earth taken by the Voyager 1 space probe from a distance of six billion kilometres. Sagan requested that the probe turn around 180 degrees after it completed its science goals and take a photograph. It shows the Earth as a 'pale blue dot' occupying a single pixel of the image. In his 1994 book 'Pale blue dot', Sagan comments on the significance of the image in five paragraphs of one of the most thought-provoking texts I have ever read. I urge you to listen to Sagan himself reciting his own words – you can find it on the internet.

Although Sagan was inspired as a child by the idea that Mars hosted abundant life, it would be Sagan himself who dashed his own hopes. In the 1950s scientists thought that the changing patterns of light and dark on Mars seen through telescopes indicated vegetation that changed with the seasons – Sagan showed that the variations were caused by Martian storms blowing dust across the surface. He was one of the first to hypothesise that the moons of Jupiter and Saturn could host oceans and life – the first idea has been confirmed, but whether these worlds do host life is yet to be tested. The European Space Agency launched a mission in 2023 to explore the moons of Jupiter and to search for signs of life that they may harbour.

In 1972 Sagan and his colleague George Mullen were the first to calculate the past and future of our star along with the implications for life on Earth. As the Sun turns mass into energy by fusing four protons together to form helium nuclei, the central core becomes less massive and has fewer particles to provide pressure to resist gravitational collapse. Consequently, the core of our star contracts and heats up as gravitational energy is converted to kinetic energy. This leads to a higher rate of nuclear reactions and the luminosity, or power output of the Sun, slowly and steadily increases over time. The higher internal temperature will cause the outer regions of the Sun to expand. Thus, as time ticks on, the Sun is getting larger and more luminous, whereas in the past it was smaller and fainter.

Sagan and his colleague calculated that when the fusion reactions at the Sun's core started 4.58 billion years ago, it shone with a luminosity that was about two thirds of today's value. He estimated that for the first two billion years of Earth's history, temperatures should have been so cold that there would have been no liquid water since the oceans should have been frozen. But the geological evidence indicates that average surface temperatures on Earth have been reasonably constant over the past few billion years despite the ever-increasing amount of sunlight it receives. And from the existence of ancient fossilised stromatolites, we know there was life and liquid water on Earth at least three and half billion years ago. This dilemma is now known as 'the faint young Sun paradox'. Sagan and Mullen proposed that in the past our Earth had a thicker atmosphere with a higher concentration of greenhouse gases which could have kept temperatures much warmer on the surface of our planet.

Sagan and Mullen then showed that as the Sun's luminosity continues its steady rise into the future there are drastic consequences for life on Earth. As temperatures on Earth increase, the carbon dioxide level in the atmosphere decreases as the rate of chemical weathering processes increase. Chemical weathering is a natural process during which atmospheric carbon dioxide combines with water to form carbonic acid. The carbon dioxide literally rains from the sky and bonds with surface rocks. In about 500 million years, the carbon dioxide levels will drop below the level needed for plants and trees to photosynthesise leading to their extinction. In one billion years from now the luminosity of the Sun will be ten percent higher and average temperatures on Earth will have risen to almost 50C. There will be a runaway greenhouse effect as the oceans evaporate into the atmosphere. By this time plate tectonics is likely to halt, Earth will resemble Venus and all animal life on Earth will come to an end. That is quite a stunning thought. Life on Earth has enjoyed 4 billion years of ideal conditions, yet we can only enjoy another 500 million years before conditions become too extreme for life like us[73]. If by some miracle our species manages to survive for this time then we may have to evacuate our planet and find a new home, but at least Sagan has given us plenty of warning so we may prepare.

During the cold war he worked with the Soviet Space Research Institute in Moscow to promote a joint US-Soviet mission to send astronauts to Mars. Sagan believed that such a joint enterprise could defuse the arms race and at the same time search for life on Mars. The collapse of the Soviet Union and the Space Shuttle Challenger disaster led to the demise of that mission. During the last years of his life, Sagan devoted much of his time to fighting for nuclear disarmament – warning the world of the nuclear winter that would ensue if a war with such weapons occurred.

Sagan received many awards, but only after the age of 50 and nearly all for his writing and popularising science. These included Emmy's for the PBS series Cosmos, the Pulitzer Prize and the Hugo Award for non-fiction. He was also selected as 'Humanist of the year' in 1981 by the American Humanist Association. Probably because of 'the Sagan effect' his research was never recognised. He certainly deserved recognition for pioneering the research area of exobiology in addition to his inspirational words. Sagan died from the effects of a rare bone

---

[73] Note that this slow increase in temperature on Earth has nothing to do with the rapid man-made global warming we are seeing today. Humans have warmed the planet by over one degree in a century, an amount that would take 20 million years for our Sun.

marrow disease aged 62. His memory will live on in the generations of scientists and the public he inspired. My favourite quote of Sagan is short but profound: *"We are a way for the cosmos to know itself"*[74].

Sagan's vision of life existing out there in the universe will probably survive for billions of years – he designed the plaques on the Pioneer probes that were launched from Earth in the early 1970s. He also designed the famous golden records that are onboard the two Voyager spacecraft. Launched in 1977 they are currently the most distant man-made objects from Earth. These lonely probes are on their way out of the solar system and will orbit within the stars of our galaxy. They will most likely never be discovered, but they will exist amongst the stars for billions of years, long after Earth is no more, as in the vacuum of space there is little to destroy them.

---

[74] Cosmos: A Personal Voyage, 1980, Carl Sagan, PBS, Episode 1 'The Shores of the Cosmic Ocean'

33. Emmy Noether (1882-1935)

"for the connection between space and time and conservation laws"

Descartes and Newton introduced the concepts that quantities such as circular or angular motion should be conserved. That means that the motion of an object in space, such as the journey of the Voyager Probes, will continue for eternity unless they hit a star, and that the rotation of our planet will continue forever unless another force acts to prevent it. That's good, otherwise the speed at which our Earth orbits might change over time causing it to spiral into the Sun or away into our galaxy! The idea that mass is conserved could be credited to the teachings of Mahavira, Jain philosophy around 500BC – that the universe and its contents, such as matter, cannot be destroyed or created. From ancient Greek philosophy also comes the idea that 'nothing comes from nothing'. In the 17th century, the German scientist Gottfried Leibniz introduced the concept of energy of motion – he called it the vis viva, or living force of a system, stating that it must be conserved. The idea that heat energy is conserved was developed in the 19th century. But it was Einstein who showed that matter and energy are equivalent - now, we simply state that mass-energy must be conserved.

But why does nature give rise to these conservation laws? They were taken as axioms, facts that were facts just because that is how nature works. It was the brilliant German mathematician, Emmy Noether who discovered the physical origin of these laws. Noether has been described by many, including Einstein, as the most important woman in the history of mathematics. But the gender distinction is not needed - Emmy Noether was one of the greatest mathematicians of all time and her most famous work had implications across all of science, particularly in physics.

Emmy Noether was born in Erlangen, Bavaria. Her father was a mathematics professor at the University of Erlangen and her mother came from a wealthy Jewish family. Initially her aim was to become a language teacher and she studied English and French, obtaining her teaching certificate in 1900. But she decided to pursue a career in mathematics. For the next few years, she attended lectures in Erlangen and in Göttingen, taught by the likes of Karl Schwarzschild, David Hilbert and Hermann Minkowski. At this time women had to obtain special permission to attend lectures. Noether was just one of two women out of nearly a thousand men at the university. In 1904 the rules were changed and women were allowed to matriculate on an equal basis to the men.

After she completed her doctorate in 1907, she worked without pay for seven years - women were generally excluded from academic positions and were not allowed to undergo the German habilitation process (a formal qualification to become a university professor). The renowned mathematician David Hilbert became impressed with her research and invited her to lecture at the University of Göttingen in 1915. This was blocked by members of the faculty because she was a woman, one member is quoted to have said *"What will our soldiers think when they return to the university and find that they are required to learn at the feet of a woman"*. Hilbert responded *"I do not see that the sex of the candidate is an argument against her admission as privatdozent. After all, we are a university, not a bath house"*[75].

Noether began teaching at Göttingen without an official position, or pay, for several years. It was during this time that she came up with a remarkable theory as to why conservation laws exist. Noether's theorem, as it is now called, revealed how any continuous symmetry of nature gives rise to a unique

---

[75] As recalled by the distinguished mathematician and colleague of Noether, Hermann Weyl, in his memorial address at Bryn Mawr College 1935, Scripta mathematica III, 3, pages 201-220.

conservation law. The formal statement of Noether's theorem, and its derivation, are rather complex and lengthy. But the basic idea is simple and profound.

A continuous symmetry means that a transformation which is performed on a physical system always leaves its measurable quantities unchanged. The symmetry of a snowflake is not continuous but discrete – it only looks the same if you rotate it by 60 degrees. The symmetry of a perfect ping-pong ball is continuous – no matter how you rotate it, it will always look the same and there is no way that you could tell that it had been rotated. The symmetry of rotation - that space is the same in all places gives rise to the conservation of angular motion.

The symmetry of translation – that space is the same in all directions, in the sense that if you are moving through space, it doesn't matter in which direction you are moving, your physical description will be the same. This leads to the conservation of momentum and the true origin of Newtons first law – a body at rest will stay at rest, a moving body will continue to move in a straight line unless acted on by a force. And most beautifully, that time is symmetric (whether flowing forwards or backwards, the laws of physics do not change) leads to the conservation of energy. Noether's theorem also gives rise to several important conservation laws in the quantum mechanical description of particles.

After the end of World War I there was a significant change in social attitudes, including more rights for women. In 1919 the University of Göttingen awarded Noether her habilitation. Three years later she received a letter from the Prussian Minister for Science, Art, and Public Education, in which he conferred on her the title of *nicht beamteter ausserordentlicher Professor* (an untenured professor with limited internal administrative rights and functions). Although it recognized the importance of her work, the position still provided no salary. Noether was not paid for her teaching until she was appointed to the special position of *Lehrbeauftragte für Algebra* a year later.

Herman Weyl summarised the mood in German after World War I in his memorial address: *"During the wild times after the Revolution of 1918, she [Noether] did not keep aloof from the political excitement, she sided more or less with the Social Democrats; without being actually in party life she participated intensely in the discussion of the political and social problems of the day. ... It is hardly imaginable nowadays how willing the young generation in Germany was at that time for a fresh start, to try to build up Germany, Europe society in general, on the foundations of reason, humaneness and justice. But alas! the mood among the academic youth soon enough veered around; in the struggles that shook Germany during the following years and which took on the form of civil war here and there, we find them mostly on the side of the reactionary and nationalistic forces. In later years Emmy Noether took no part in matters political. She always remained, however, a convinced pacifist, a stand which she held very important and serious."*

In the following years Noether made many contributions to mathematics, in particular algebra and topology - highly technical works that were, and still are, highly regarded by the world's most prominent mathematicians. Her reputation became widespread and she was twice invited to address the International Congress of Mathematicians. Following her address to the 1932 Congress in Zurich she received the Alfred Ackermann-Teubner Memorial Prize for the Advancement of Mathematical Knowledge.

Later, in 1933, she was one of the numerous academics at German universities that were dismissed because of their Jewish ancestry. Weyl summarised the rapid decline of the world's greatest centre of mathematics research: *"In the spring of 1933 the storm of the National Revolution broke over Germany. The Göttinger Mathematisch-Naturwissenschaftliche Fakultät, for the building up and consolidation of which Klein and Hilbert had worked for decades, was struck at its roots. ... Emmy Noether, as well as many others, was prohibited from participation in all academic activities, and finally her venia legendi* [teaching permission], *as well as her Lehrauftrag* [teaching assignment] *and the salary going with it, were withdrawn. A stormy time of struggle like this one we spent in Göttingen in the summer of 1933 draws people closer together; thus I have got a particularly vivid recollection of these months. Emmy Noether, her courage, her frankness, her unconcern about her own fate, her conciliatory spirit, were, in the midst of all the hatred and meanness, despair and sorrow surrounding us, a moral solace. It was attempted, of course, to influence the Ministerium and other responsible and irresponsible but powerful bodies so that her position might be saved. I suppose there could hardly have been in any other case such a pile of enthusiastic testimonials filed with the Ministerium as was sent in on her behalf. At that time we really fought; there was still hope left that the worst could be warded off. It was in vain. ... It shall not be forgotten what America did during these last two stressful years for Emmy Noether and for German science in general."* As Weyl wrote these words in 1935, he would not realise the true extent of the horrors that would be coming a decade later

Noether's brother, Fritz, was also an accomplished mathematician. Not allowed to work in Nazi Germany he moved to the Soviet Union to be a professor at the University of Tomsk. In 1937 he was arrested during Stalin's Great Purge.

He was accused of being a German spy and sentenced to 25 years in prison. Fritz served several years before he was accused again of anti-Soviet propaganda and executed by firing squad in 1941.

Noether was more fortunate – she was considering moving to Moscow, but thanks to a grant from the Rockefeller Foundation she took a position at Bryn Mawr College in America. She enjoyed the last two years of her life in the United States lecturing on mathematics and training a new generation of female mathematicians, until she died suddenly in 1935 following surgery for a tumour aged just 53. Albert Einstein wrote in the New York Times: *"Within the past few days a distinguished mathematician, Professor Emmy Noether, formerly connected with the University of Goettingen and for the past two years at Bryn Mawr College, died in her fifty-third year. In the judgment of the most competent living mathematicians, Fraulein Noether was the most significant creative mathematical genius thus far produced since the higher education of women began. In the realm of algebra, in which the most gifted mathematicians have been busy for centuries, she discovered methods which have proved of enormous importance in the development of the present-day younger generation of mathematicians."*

Because of her gender, Emmy Noether never held a permanent position at a university. There is no Nobel Prize for mathematics, instead, the highest award for such research is the Fields Medal which was established in 1936, the year after Noether died. She would have surely deserved this award for her contributions to mathematics. Her discovery of the reason for conservation laws transcends mathematics and has played a crucial role in the development of our understanding of the workings of the universe – for that, she certainly deserved the Nobel Prize in physics.

34. Chien-Shiung Wu (1912-1997)

"for showing that nature is not always symmetric"

One of the unsolved problems in our quest to understand our universe is the question of why there is any matter at all? It was long thought that the laws of physics were always symmetric in space and time, in which case the natural outcome from the early universe would have been an equal amount of both matter and their antimatter particles created during the first second of our universe. Nearly all of those particles would have collided and annihilated into photons and neutrinos, leaving nothing to form stars or life. An old idea was that some mechanism separated the matter from the antimatter, resulting in large regions of our universe being made of one type of matter. But this would leave an observable signal at the boundaries of such regions where matter and antimatter come into contact, and that has not been detected. Another possible solution is if interactions between particles were not symmetric, giving rise to an excess of matter over antimatter. That nature sometimes violates symmetry, leaving room for a solution to the origin of an excess of matter in the universe, was first discovered by the Chinese experimental physicist Chien-Shiung Wu. She showed that a phenomenon and its mirror image are not always the same, that the universe does at times distinguish between left and right.

Wu was born in China in 1912 when the New Cultural Movement was just beginning. The Confucian beliefs of the importance of men and their dominance over women was ending. It was the year that the centuries old practice of foot-binding women was officially outlawed. Wu's parents played a role in the women's liberation movement that encouraged women's equality and the education of Chinese women. Her father took part in the revolution of Sun Zhongshan that led to this modernisation of China. In 1919 the first independent women's university run by the Chinese government was founded in Beijing. It was the beginning, but progress was slow – it only really affected a small number of elite women, the vast majority of women in the countryside were unaffected by these changes. It was not until 1949 when Mao Zedong abolished the feudal system and stated *"women hold up half the sky"* that women's rights in China began to undergo a widespread change.

Wu was educated at a school for girls that was founded by her father and she later majored in physics, coming top of the class at the National Central University (now Nanjing University). Wu then worked as an assistant to Gu Jung-Wei, a female professor of physics who had earned her PhD at the University of Michigan and advised Wu to do the same. Beginning with the New Cultural Movement, young elite Chinese females began to make the journey to America for education – such privileges were unthinkable for most women in China. But despite being elite, they were pioneers in the Chinese women's liberation movement.

Wu applied to Michigan University and was accepted. Her travel funded by a wealthy uncle, she arrived by boat in San Francisco in 1936 and visited the laboratories at Berkeley where she met Ernest Lawrence, the director of the nuclear research laboratories. Lawrence received the 1939 Nobel Prize in physics for inventing the cyclotron particle accelerator. Lawrence was so impressed after talking with Wu that he encouraged her to change her plans and offered her a PhD position. At the same time, she learned that females were not allowed to enter in the front entrance to Michigan University student's union so she decided to study at the more liberal University of California at Berkeley.

Wu quickly became known as a careful and brilliant experimental physicist. She graduated in 1940 after carrying out research on nuclear fission and soon after became involved in the Manhattan project due to her expertise. After the war she accepted an associate research professorship at Columbia University. There she carried out experiments on quantum entanglement and precision experiments of beta decay that confirmed Enrico Fermi's new theory of the weak interaction. In 1956 she had planned to travel back to China with her husband when her Columbia colleague Tsung-Dao Lee discussed a crazy idea with her, one that he had formulated with his colleague Chen Ning Yang. The idea began with a suggestion from Richard Feynman, that certain interactions between particles might violate the widely assumed symmetric laws of nature.

We have already heard about the postulate of Poincaré, that the laws of physics should be unchanged in space and time. That means the physics over there is the same as over here, that physics is the same regardless of whether time flows forwards or backwards. We have also heard about Emmy Noether's work on symmetries and conservation laws. But to produce an excess of matter over antimatter some symmetry needs to be broken. Such a possibility was declared as nonsense by the famous particle physicist Wolfgang Pauli. Much to his, and every theoretical particle physicist's surprise, Wu proved that sometimes nature violates our expectations.

Imagine looking at a clock in a mirror – it appears to turn counter-clockwise. Imagine constructing a clock such that each component was a mirror image of itself. The screw threads would become right-handed instead of left-handed, the numbers would be printed as their mirror images. The clock hands would turn counter-clockwise.

Some particles, such as protons and neutrons are themselves made up of smaller particles called quarks. Quarks and electrons are described as fundamental particles – that is they are not made of smaller entities. We can't explain the physical appearance of a quark or an electron, they are defined only mathematically, and each has several properties such as charge, mass or spin. It is the particle's spin that gives it a 'handedness' or chirality. The particles are not really spinning, the word spin describes the quantum mechanical counterpart to angular momentum. Now consider a neutron which can decay into a proton with the emission of an electron and an anti-neutrino. This is an example of a process called beta decay and it operates under the weak force. If we made a mirror image of the particles (by flipping the spin of the proton for example), the mirror image of the reactions should take place. But it doesn't.

Wu looked at the decay of cobalt to nickel during which one of the cobalt neutrons decays to a proton by emitting an electron and an anti-neutrino. She first supercooled the cobalt to 0.01 degrees above absolute zero to reduce their thermal motions, and then applied a strong magnetic field which causes the spin of all the cobalt nuclei to align. This is important since this spin axis should be imprinted on the new particles which emerge from the decay. Wu measured the direction in which the electrons emerged. She then repeated the experiment changing the direction of the magnetic field, flipping the spins of the nuclei. If nature were symmetric, she would have found that aligning the nuclei spins in one direction would produce identical results to aligning them in the other direction. Wu found that the direction of the emerging electrons was asymmetric – the electrons emerged in a direction that were preferentially opposite to the nuclei spins. Wu had discovered what we call a violation of parity symmetry confirming that the weak interaction violated mirror symmetry and that nature differentiates between left and right.

Imagine sitting on a rotating chair and dropping a ball. It falls downwards. If you viewed yourself in a mirror, you would appear to be spinning the opposite way but the ball would still appear to fall downwards. The surprise in Wu's experiment was the equivalent of the ball falling upwards!

After Wu announced her results The New York Times heralded the *"shattering of a fundamental concept of nuclear physics"* on its front page. Soon after her discovery, many other experiments were performed that revealed the same effect in different forms of the weak interaction.

Wu's experiment was the first important step towards answering the question posed at the beginning of the chapter – "why does matter exist in our universe?" She showed that the weak interaction violates parity symmetry – that the rules change when you reflect the system in a mirror. In 1964 another symmetry of nature was shown to be broken – so called charge-parity[76]. In this case symmetries are broken when particles are replaced with their anti-particles, and antiparticles are replaced with their particles.

In 1967 the Russian physicist Andrei Sakharov described three necessary requirements for how nature must operate in order that an excess of matter over antimatter is produced during the big bang. In addition to the symmetries discussed above, there is one more symmetry that needs to be violated, that of the so-called baryon number. This has not yet been observed in an experiment, but no doubt if discovered then

---

[76] The experimentalists who discovered this, James Cronin and Val Fitch, were awarded the Nobel Prize in physics in 1980.

more Nobel Prizes be deserved. And even if this problem is solved, explaining why there is an excess of matter over antimatter, it does not answer the question of why there is something and not nothing.

In 1958 Wu became the first female professor in the history of Columbia University. She continued carrying out experiments that helped establish the standard model for particle interactions. She would later receive the National Medal of Science, the Wolf Prize in physics and over a dozen honorary doctorates. For their prediction that parity might be violated, Tsung-Dao Lee and Chen Ning Yang were nominated once for the Nobel Prize in 1957 and received the prize in 1958. Between 1958 and 1965 Wu was nominated seven times for a Nobel Prize for her discovery. The archive is not available after this time. That she was not included in the 1958 prize award is yet another of the big oversights of the Nobel Prize selection committee.

In her later life Wu spoke out against gender discrimination and human rights, long after she retired in 1981. She died of a stroke in 1997 in New York, and her ashes are buried in the courtyard of the Ming De School that her father had founded and where she learned her love of physics.

35. Edward Tryon (1940-2019)

"for the idea that the universe could begin from nothing via a quantum fluctuation"

Why is there something and not nothing? This is the biggest question of them all, and one that many have asked. That *"nothing comes from nothing"* was stated by the ancient Greek philosopher Parmenides in the 5th century BC. This question remained in the realm of philosophy and theology for two and half thousand years. It was considered beyond the possibility of answering by physicists and rarely discussed. The only widely accepted solution was that an all-powerful being that existed for eternity created the universe. That is, until the work of Edward Tryon, a physicist that very few of my colleagues have even heard of.

Edward Tryon was bold enough to try to answer the question and came up with the grandest idea of all. He published his theory in the journal Nature in 1973 titled 'Is the universe a vacuum fluctuation?'

That something cannot arise from nothing is a fundamental consequence of the conservation laws of mass and energy. Since the universe is 'something', most physicists thought it was hopeless to try to come up with a mechanism for its origin from nothing. Tryon starts out by pointing out that a universe with particular properties could appear from nowhere – one in which the total energy is zero. In such a universe, spacetime would be flat[77]. He goes on to discuss quantum field theory, where in principle, every event that can happen will happen. A vacuum that is apparently devoid of particles is actually a foaming sea of activity and fluctuating energy. It can be described mathematically with particles and their antiparticles constantly popping in and out of existence on timescales that

---

[77] I believe that this was first pointed out by the American physicist Richard Tolman in his 1934 monograph 'Relativity, thermodynamics and cosmology'

are unmeasurably small. Tryon argues that there is no limit on the scale of such vacuum fluctuations, and that our universe could arise from a simple quantum fluctuation in a vacuum.

Tryon was a modest scientist, in his paper on the origin of a universe from nothing, he states *"In answer to the question of why it happened, I offer the modest proposal that our universe is simply one of those things which happen from time to time."*[78] He mentions that the vacuum must initially exist within a larger space in which our observable universe was embedded. Tryon is foreseeing the idea of a multiverse. He then discusses the role of life in such a universe – one of the first 'anthropic arguments' ever made. That any universe in which sentient beings find themselves is necessarily hospitable to sentient beings. He uses this argument to suggest that even though universes that emerge from such spontaneous fluctuations are probably rare, observers of a universe will only evolve in those universes that have conditions suitable for life.

Tryon studied at Cornel University where he was influenced by Nobel Laureate Hans Bethe. He then went on to the University of California at Berkeley and was mentored for his PhD by Nobel Laureate Stephen Weinberg. His thesis work was on the relationship between general relativity and quantum field theory and in 1971 he took up a teaching position at the City University of New York. When he sent his paper on the origin of the universe to the journal Physical Review Letters, they declined to publish it. He then sent it to the more prestigious journal Nature where they made it a feature article. Despite that, and being reported in the newspapers, it was subsequently forgotten by the scientific community.

---

[78] Edward Tryon, 'Is the universe a vacuum fluctuation', Nature, Vol 396, 1973, page 397.

The idea of a universe emerging from nothing picked up again in 1982 after the theory of inflation became popular. The Ukrainian cosmologist Alexander Vilenkin wrote a paper describing in more mathematical detail how a universe could emerge from a quantum fluctuation within the framework of the newly proposed inflationary scenario. His paper was titled 'Creation of universes from nothing', yet the work of Tryon was only mentioned in a brief note added just before publication[79].

In the 1970s the energy of our universe was thought to arise from two components. The positive energy of the matter and the negative energy of gravity. When astronomers tried to measure the energy-density of the universe they first came up with numbers that had a net positive energy. The universe was going to expand forever. Then came the discovery of 'dark energy', in essence the cosmological constant that Einstein had discarded. It acts like a negative pressure, the opposite of gravity, and it causes the universe to expand faster and faster. However, when this component of energy is included, and if our interpretation of energy is correct on a cosmic scale, the total energy of our universe is indeed zero. This is supported by observations of the cosmic microwave background fluctuations which show that spacetime has a geometry which is very close to flat[80].

Tryon's speculation was correct! Our universe could have begun with zero energy, you just had to take a piece of spacetime and fill it with positive and negative energy such that their sum was nothing but it began expanding. But that is

---

[79] Often such a footnote indicates that the author came up with the idea independently and an anonymous referee pointed out that the idea was not original.

[80] See Appendix II for an overview of spacetime.

not nothing, it is something, even if it is just an empty piece of spacetime. Tryon has not completely answered the question of why our universe emerged from nothing, but the idea is so profound that he deserves the recognition.

What is nothing anyway? There is nowhere in our universe where you can find nothing. Any random few cubic metres of space will contain a lot of 'stuff'. Perhaps a proton, a few dark matter particles and certainly a vast number of photons and neutrinos. What if you found a special small volume of space with none of these particles, is that what nothing is? It would still contain dark energy. It would still be a volume of our spacetime. It would still undergo quantum fluctuations. Our physical laws would still exist within that empty volume – as I mentioned much earlier, our most fundamental physical laws, of conservation of energy and motion, they come only from the symmetry of space and time. But if you remove space and time, or spacetime, then these laws do not exist.

Perhaps nothing is what is left when you take away the entire universe with all of its physical laws? In the context of physics it is impossible to make sense of absolute nothingness. We do not have a physical theory to describe what happens outside of the universe and beyond the realm of its physical laws. The concept of nothingness is physically ill-defined. Despite this, some cosmologists have been so bold as to state *"'nothing' is unstable"* and will naturally result in the spontaneous appearance of quantum fields.

We were taught at school that particles are the building blocks of everything. But that isn't correct, the underlying building-blocks are 'quantum fields', so let's dissect a quantum field. In physics, a field is a set of values that exist at all points in space. The temperature measured at all points in a room could be considered a 'temperature field'. A quantum field encodes the laws of quantum mechanics into a field that has values at all points in space. This means that the energies within that field are 'quantised' and can only take on certain values. This is quite different from our temperature field in which the temperature can be any numerical value.

A quantum field describes how forces propagate and how particles manifest themselves. Quantum fields are not a physical object that I can describe, but a mathematical framework that works extremely well. I envisage a quantum field as an ether-like energy, spread throughout all of space. The ripples in this energy field give rise to electromagnetic waves and light. Every particle that exists, from electrons to quarks, and all our physical laws and forces, are manifestations within an underlying quantum field. Even if we look at a volume with no particles, a perfect vacuum that we call a quantum vacuum, the quantum field is still present. And within that empty volume, we can interpret the mathematics as describing a sea of particles and anti-particles that are constantly being created and destroyed.

The bold but speculative string theory attempts to create a deeper understanding of such quantum fields. It replaces the point-like particle manifestations of the quantum field with tiny one-dimensional vibrating strings. The vibrational states of the string leads to the properties of particles that we measure, such as mass and charge. These objects exist in a ten- or eleven-dimensional space that is curled up so small that we do not detect its presence. Unfortunately, there is not a clear experimental test that could be carried out to determine the veracity of such theories.

We can reduce our grand question to why there was a piece of spacetime, a quantum vacuum that may have randomly fluctuated to create our entire universe. That in itself is a profound simplification of the question. But we are still left with the question of why that quantum vacuum existed? Of course, you can still speculate that it was so designed and created by a supernatural being. If that idea gives you comfort, then fine. We cannot exclude that, but it does not answer our question. A supernatural being smart enough to create a piece of spacetime with a quantum fluctuation from nothing is certainly not nothing. It is a far more complex solution than the idea that there was an eternally existing empty spacetime with a zero-energy state which happened to undergo a chance fluctuation. That the universe just suddenly appeared without a reason. Many scientists are happy to accept that the quantum mechanical view of the world does not necessarily require 'cause and effect'. In the quantum mechanical world things can happen spontaneously and randomly, or at least they appear to do so. I think that would be a very unsatisfactory description of how things behave, but it may be one with which we have to live.

I wish I could tell you more about Edward Tryon, but little is known of his life. There are no biographies of Tryon. You will not find a single image of him on the internet. His work is rarely mentioned in books on the origin of the universe. In the popular science book devoted to this question, 'A universe from nothing' by Lawrence Krauss, Tryon is not mentioned despite his ideas being the focus of the book. Neither is he mentioned in Stephen Hawking's 'A brief history of time' which also features his ideas. Tryon passed away in 2019 after suffering from Alzheimer's disease. His brief obituary in the New York Times tells of his work on a universe from nothing and mentions that he enjoyed fishing and was very fond of cats, often taking in strays from his neighbourhood.

36. Ludwig Boltzmann (1844-1906)

"for the interpretation of thermodynamics and the arrow of time"

No scientist likes their work to be unrecognised. Ludwig Boltzmann struggled to obtain recognition for his theories, that the behaviour of matter could be understood by treating it as if it were made of atoms. Some writers have speculated that this may have contributed to Boltzmann ending his own life in 1906. Despite his research being over a century old and predating the discovery of quantum mechanics and the big bang, his ideas were revolutionary and it is not surprising that their importance was misunderstood. Today, Boltzmann's research is still impacting current research on fundamental questions such as the nature of time and the beginning and end of the universe.

Boltzmann was born in Austria. His father, a custom official, died when he was 15 and his mother used her small inheritance to enable her son to attend university. He began studying mathematics and physics at the University of Vienna in 1863. He quickly received recognition for his fundamental research and his inspirational lectures, and by 1869 he was already appointed to full professor at the University of Graz.

In the 19th century the laws that described the behaviour of matter, how heat and temperature are related to energy and entropy, were developed under the topic of thermodynamics. These were empirical laws discovered through experiments that were trying to improve the efficiency of steam engines. The first law of thermodynamics is a statement about conservation of energy – that energy can only be transformed and never destroyed. The second law of thermodynamics, that the entropy of a closed system can only increase, is more complex to understand, but equally fundamental. It explains why perpetual motion machines cannot exist, and ultimately, it governs our fate in the far future of the universe.

The second law of thermodynamics resulted from the work of the French scientist Sadi Carnot and the German scientist Rudolph Clausius in the middle of the 19th century. That entropy always increases stems from the observation that heat always flows from hot to colder objects, but in the process some of the 'useful heat' becomes lost. The lost heat was characterised by a quantity called entropy. At the time, many scientists thought that heat was a weightless fluid that flowed between objects. It was only by the beginning of the 20th century that heat was accepted to be a consequence of the kinetic motions of molecules. We now know that some of the heat energy is radiated away and some is lost by conduction leading to an increase in the kinetic motions of the surrounding air. The second law of thermodynamics states that the amount of useful heat that can be used to do work, is constantly decreasing.

Many scientists thought that the laws of thermodynamics were the end of the story and that there was no deeper explanation, it was just matter behaving as matter. Boltzmann showed how the laws of thermodynamics arose from the collective behaviour of atoms, before atoms were even proven to exist. Boltzmann began by showing how collisions between atoms in a gas lead to a characteristic distribution of speeds. This is now called the 'Maxwell-Boltzmann distribution', since its form was guessed by the Scottish physicist James Maxwell and derived by Ludwig Boltzmann. It is used pretty much every time a gas is modelled, from Earth's atmosphere to the early universe plasma. By 1872 Boltzmann had derived his famous 'Boltzmann transport equation' that describes how a collection of particles evolves in time. It is an equation used today to study many phenomena, from the evolution of stars in a galaxy to the cosmic microwave background radiation.

Boltzmann went on to derive his famous 'H-theorem' that he believed proved the second law of thermodynamics based on the atomic treatment of matter.

His radical ideas were met with a lot of scepticism, particularly by scientists such as Ernst Mach and Wilhelm Ostwald, who argued that there was no evidence that atoms existed, or that heat should treated as a fluid. But the most problematic attack on his theories came from his friend and Austrian colleague Josef Loschmidt.

Loschmidt argued that every known law of physics works in the same way if you reverse time. That physics works just as well backwards in time as forwards in time, therefore how can it lead to a quantity – entropy - that only increases in one direction of time? As an example, you can imagine watching a pendulum and reversing time. It would still look like a swinging pendulum that can be described by simple physical laws – you would not be able to tell that time was flowing backwards. Loschmidt's argument was a valid observation that Boltzmann struggled with until the end of his life, and it is still not completely resolved today.

Boltzmann countered the arguments of Loschmidt, and in 1877 he showed that the entropy of a system is proportional to the logarithm of the number of possible states (such as position and velocity) of its constituents. He argued that an initial well-ordered state of molecules would become increasingly disordered - that the second law of thermodynamics was simply the result of the fact that such disordered states were the most probable and that nature evolves towards the most probable state. And because the number of possible disordered configurations of the molecules is much higher than organised configurations, a system will almost always be found in the state of maximum disorder. It is the reason that an egg is easy to break, but virtually impossible to put back together. It explains why when you stir cream into coffee it becomes perfectly mixed, but if you reverse the direction of your spoon, it will not unmix. It is why the scent of a candle will fill a room and will not suddenly become concentrated again inside the candle instead of the whole room.

Boltzmann showed that entropy 'nearly' always increases, rather than strictly always increases. This was the first time that probabilities had ever been used within a so-called fundamental law of physics, breaking the centuries old idea that the laws of nature were deterministic. Boltzmann's counter argument to Loschmidt would have been that the difference between the pendulum and the broken egg, is that the pendulum is just one body, whereas the egg is made of many trillions of molecules. That the molecular chaos of numerous interacting bodies makes the egg virtually impossibly to put back together. Boltzmann was correct in his counter arguments but it still did not completely solve the conundrum raised by Loschmidt.

This led Boltzmann to ponder the nature of time. In his 'Lectures on gas theory' from 1897 he remarked: *"just as at a particular place on the earth's surface we call 'down' the direction toward the centre of the earth, so will a living being in a particular time interval of such a single world distinguish the direction of time toward the less probably state from the opposite direction (the former toward the past, the latter toward the future)"*[81]. In other words, Boltzmann associated the passage of time with the direction at which entropy must increase. That time proceeds in one

---

[81] Boltzmann L. In: Lectures on Gas Theory. Brush S., translator. University of California Press; Berkeley, CA, USA: 1964, pages 402-403.

direction is obvious to us, because we remember the past and not the future. But the physical interpretation of 'what is time' is still unresolved today. However, Boltzmann's connection between the steady increase of entropy with the incessant passage of time is the foundation of many of today's theories of the nature of time.

The next attack on Boltzmann was along the same lines as that made by Loschmidt. In 1890 Poincaré derived his famous 'recurrence theorem' which states that after some period of time, dynamical systems like a gas, will always return to a configuration that matches any state in its past. In other words, the smoke of a candle that fills a room, must at some point return to be concentrated in the candle! The mathematician Ernst Zermelo used the recurrence theorem to criticise Boltzmann's ideas. Indeed, Poincaré's math was correct, but Boltzmann countered by showing that the timescale for a system to return to its initial ordered state was incredibly long. Boltzmann's probabilistic treatment of matter doesn't exclude such events happening, that a broken egg will spontaneously reassemble, it is just extremely improbable that it will.

The arguments then turned to the entire universe. Soon after the second law of thermodynamics was discovered, William Thomson (Lord Kelvin) observed that if you apply it to the universe as a whole, it must undergo a so called 'heat death' in the future. It follows on from the second law that two objects at different temperatures will transfer heat until they come to equilibrium. When everything in the universe is at the same temperature, entropy will have reached its maximum value and no more useful energy can be extracted from anything. Decades before the expanding universe was discovered, Thomson made the obvious but quite profound observation that the universe could not be infinitely old since the heat death has not yet happened.

Boltzmann proposed solutions for our existence within such a universe. He stated that our visible universe must have begun in a low entropy state, thus the progression of time towards its demise in a high entropy state – the so called 'heat death' of our universe. He proposed a mechanism by which the low entropy state could occur by chance, due to random fluctuations of patches of a high entropy universe returning matter to a low entropy state. In his same lectures on gas theory mentioned above he wrote: *"This method seems to me to be the only way in which one can understand the second law—the heat death of each single world—without a unidirectional change of the entire universe from a definite initial state to a final state."*

That statistical fluctuations can return a high entropy state to a low entropy state is certainly possible, but the timescales for this are vast. Today, we know from observations of the cosmic microwave background that the early universe was indeed a low entropy state – a smoothly distributed sea of matter and radiation. As time passes gravity proceeds to form structures in the universe, stars, planets, and life. But why it was in such a state is not known. Theories such as inflation and the multiverse that I will describe in the last chapters may provide part of the answer.

Boltzmann's writings spanned many topics. He used the concept of negentropy (negative entropy) to explain the nature of life: *"The general struggle for existence of living beings is therefore not a fight for the elements"* which he argued were abundant, but *"Rather, it is a struggle for entropy that becomes available through the flow of energy from the hot Sun to the cold*

Earth."[82] That living things seemingly defy the second law of thermodynamics by becoming more orderly, is because they are not isolated systems. Such idealised configurations do not exist in our interconnected universe. Life interfaces with its environment and as living things become ordered, as a seed germinates into a plant or as we create a memory, the entropy of the universe increases as a consequence.

Boltzmann's work marked the transition from classical physics to quantum physics. He foresaw the ideas in quantum mechanics by quantising energy long before Max Planck. His work influenced Planck, Einstein and many of the scientists mentioned in this book. He also wrote and lectured extensively on philosophy. This work was so well regarded that he was offered and accepted two professorial chairs at the University of Vienna, one in physics and one in philosophy.

Boltzmann was a passionate man, described as a bulldog in debates by his colleagues, he would openly weep at the train station when he had to leave his wife for a long period of time. He suffered from bouts of depression – sometimes amongst guests, he would lapse into lengthy silences from which he could not be drawn. He described how his mood would swing from intense joy to deep grief. Later in his life Boltzmann suffered from heavy attacks of asthma and headaches, and his eyesight suffered from myopia. Despite his declining health, he still travelled to America in 1904 and 1905, writing a particularly humorous essay on his travels to California titled 'Journey of a German Professor to Eldorado' that was published in his 'Popular Writings'. By 1906 his mental condition had declined such that he could not complete his

---

[82] Der zweite Hauptsatz der mechanischen Waermetheorie. Lecture presented at the 'Festive Session' of the Imperial Academy of Sciences in Vienna, May 29, 1886.

university lectures. He committed suicide in 1906 whilst on vacation in Trieste with his family. Before his death in 1906 Boltzmann was nominated five times for a Nobel Prize in physics, twice by Nobel laureate Max Planck. He certainly deserved that honour.

Some writers have put Boltzmann's suicide down to the lack of acceptance of his theories and the continual attacks from his colleagues. This is made more dramatic since the year after his death evidence for atoms was discovered. Others have pointed out that Boltzmann was highly regarded and famous worldwide - he was offered numerous prestigious professorships and held positions at many universities in Germany and Austria – and there was the suggestion that his suicide was due to his manic depression. Perhaps it was a bit of both. Being famous does not prevent people from taking their own lives, it can be a contributing cause, particularly if they feel ignored even if they are not.

Boltzmann's fame will live on in his theorems and methods, and the many research institutes and foundations that bear his name. And as we still try to uncover the meaning of time and the origin and destiny of our universe, his name is currently mentioned in the titles of over 1000 scientific publications each year. His famous formula for entropy, $S = k \ln W$, is engraved on his tombstone in Vienna.

37. Pier Giorgio Merli (1943-2008), Gian Franco Missiroli (1945-) & Giulio Pozzi (1945-)

"for the most beautiful experiment ever performed"

In a 2002 poll conducted by Physics World, readers were asked to name the most beautiful experiment in physics ever performed. Amongst the top answers were several of the experiments I have already mentioned: Galileo's experiments with falling objects, Newton's decomposition of sunlight with a prism, Eratosthenes' measurement of the size of the Earth and Cavendish's torsion beam experiment. But top of the list was the 'single electron interference experiment'. Whenever you read a book or research study on quantum mechanics, you will inevitably come across this experiment as capturing the essence and mysteries of the theory. It is mind boggling for anyone who reads about it for the first time. What is less known, is who performed this experiment first.

The single electron interference experiment was carried out by three Italian physicists in 1974. Giuli Pozzi and Gian Franco Missiroli were professors of physics at the University of Bologna, Pier Giorgio Merli was President of the Italian Society of Electron Microscopy from 1984 to 1987 and worked at the LAMEL-CNR Institute in Bologna. Missiroli was passionate about education and novel ways to teach physics, which led him to initiate an experiment that would elucidate the mysteries of quantum mechanics.

Quantum mechanics is a mathematical description of the behaviour of particles, from atoms to photons. It arose from the failure of classical theories of matter and light to explain several observations. Some experiments, beginning with Thomas Young's famous double slit experiment at the beginning of the 19th century, showed that light must behave like a wave as it travels through space. Other experiments, such as the photoelectric effect – the emission of electrons when light hits a material - required light to behave like a particle.

Many scientists worked on developing the theoretical framework of quantum mechanics, which requires quantities such as energy and momentum to have discrete values and there are strict limits on how accurately the value of these quantities can be predicted before they are measured. Whilst at the University of Zurich from 1921-1927 Erwin Schrödinger derived his famous 'wave function' which provides information on the probability of measuring any of these quantities. It means that you cannot predict with certainty what will happen, but can only assign a probability to it. The wave function that describes an electron, or a human, describes to each point in space a probability amplitude. An electron, or a human, is essentially a cloud in probability space until it is observed or measured. Schrödinger received the Nobel Prize in 1933 for his work on quantum mechanics.

It is this probabilistic aspect of nature that allows particles to 'quantum tunnel' into regions they would not be allowed to exist according to the rules of classical physics. It also implies the strange wave-particle duality of matter. That particles, molecules or light, moving through space cannot be described by a particle or a wave alone. They are a strange mixture of the two simultaneously. This defies all of our physical experiences of the world around us. Yet the theory of quantum mechanics is precise, verified to incredible precision, and has been confirmed by every experiment ever carried out to test it.

That an electron is a particle, a tiny point like object with mass and energy, can be clearly seen in a cloud chamber experiment. This is a sealed flask of vapour used as a particle detector. When a proton or electron with high energy moves through the chamber, a visible streak of its point like path can be seen by eye. But fire a beam of particles at a double slit, and you will witness their wave like behaviour. Richard Feynman called the double-slit experiment *"impossible, absolutely impossible, to explain in any classical way,"* and said it *"has in it the heart of quantum mechanics. In reality, it contains the only mystery."*[83]

The double slit is just that, a piece of opaque material with two narrow slits close together. If you shine a beam of light at the slits and look at the pattern on the wall behind you will see a series of many bands of light and dark. In the classical picture of particles, if you fire a beam of particles towards the slits, those that pass through the slits will continue to travel in a straight line to hit the detector. You should just see two bands where they impact the wall or detector. You should not see numerous bands where the particles hit the wall, beyond the regions where they could possibly land.

This is perfectly well explained if light is a wave. As the waves of light emerge from each slit, they spread out and interfere with each other. When the waves collide and are synchronised, they add up in brightness. When they are out of synchronisation by half a wavelength, they cancel each other out. This is the experiment that the English scientist Thomas Young carried out in 1801 to disprove Isaac Newton's corpuscular theory of light. You can witness the analogous effect yourself by dropping two small stones into a still pond and watching the waves interfere as they travel cross the surface of the water.

But Isaac Newton was not wrong and neither was Thomas Young. If you put a detector behind the two slits, the photons of light are found to hit the screen at discrete points, not spread out across the detector. It is as if when the photons are travelling, they behave as a wave, but when they hit the

---

[83] The Feynman Lectures on Physics, 1963, Volume III, chapter 1, page 1.

detector, such as a photographic plate, they become localised packets of energy. The same experiment was performed with a beam of electrons in 1927, proving that as expected from quantum theory, particles behave like waves just like light, until they hit the detector or wall behind the slits.

To elucidate the real mystery of quantum mechanics, in 1965 Richard Feynman described a thought experiment in his famous Feynman lectures on physics. He said that if electrons were sent one at a time towards a double slit, and you could measure where they land with a detector, an interference patter would still emerge. Now things are getting really bizarre, if a single electron is fired towards the slit, it has a physical size that is far smaller than the slit so it can easily pass through one slit. How can it pass through both slits simultaneously, such that it interferes with itself? Surely you're joking Mr Feynman?![84]

Feynman also stated that such an experiment would be impossible to perform because of its complexity. Pier Giorgio Merli, Gian Franco Missiroli and Giulio Pozzi decided to ignore the advice of Feynman and construct an experiment to verify this most bizarre aspect of nature. In 1974 they developed the equipment necessary to fire single electrons one at a time towards a device that would cause the electron to interfere with itself[85]. They also developed a detector that

---

[84] 'Surely you're joking, Mr Feynman' is the title of an edited collection of reminiscences by Richard Feynman, one of the founders of quantum electrodynamics and the path integral formalism of quantum mechanics.

[85] Although the interference device they used was not a double slit, it was an electron biprism, it has exactly the same effect. The first single-electron experiment to use an actual double slit was performed later by Pozzi and colleagues.

could record the arrival and positions of the electrons one at a time which they could display on a television monitor. When they switched on the experiment, they observed single dots on the monitor as each electron hit the detector. Over time the full image would build up – a series of bands of dots with nothing detected in between. Somehow, an individual electron passes through both slits simultaneously, interferes with itself and then becomes a particle when it is detected. Mind blowing stuff!

Merli, Missiroli and Pozzi published their results in 1976 in the American Journal of Physics titled 'On the statistical aspect of electron interference phenomena'. They also produced a short movie of their experiment which they sent to the Seventh International Scientific and Technical Cinema Festival in Brussels in 1976. It won first prize in the physics section. It was indeed, the most beautiful and thought-provoking experiment ever carried out.

What about placing a detector at one of the slits so that you can measure which slit the particle goes through? This proved difficult. If you want to detect a photon or electron you naturally have to have some interaction with matter which stops it in its tracks. Experiments that do try to infer which slit a particle passes through, inevitably identifies its 'particle nature' and then the interference pattern vanishes. And it doesn't matter what you send towards the slits, a photon, an atom, or even a cat[86]. That's right, if you could set up an experiment where you fired cats at a double slit, you would find that each cat would somehow pass through both slits, interfering with itself as if it were a wave, such that it hits the

---

[86] This is a homage to Erwin Schrödinger and his famous imaginary cat experiment which you will hear more about in the next chapter.

wall at a position that could only be described by a probability wave passing through both slits. Only when you look at the wall would the cat actually become a cat. If you looked at the cat anytime on its journey you would see it as a cat, but then the interference pattern would be lost. Although setting up an experiment to prove that with cats is currently impossible, and inhumane, it has been confirmed with molecules that individually comprise of 2000 atoms!

The experiments of Merli, Missiroli and Pozzi did not uncover anything new about quantum mechanics. The results were predicted to occur by the theory. But they were the first to show this remarkable behaviour of single particles and demonstrate in a simple way the bizarre complexity of the universe around us. No other experiment has captured the essence of quantum mechanics in such a simple way, and has captured the imaginations of both scientists and non-scientists alike.

The most common explanation of the single electron double slit experiment is that the electron travels through space as a wave and becomes a particle when it is observed. Moreover, we cannot predict exactly where it is as it travels, we can just assign a probability to its location. That everything around us is described by a probabilistic theory that we cannot comprehend is dissatisfying to say the least. This has led to another famous quote by Richard Feynman *"I think I can safely say that nobody really understands quantum mechanics"*[87]. That is, apart from Hugh Everett, who gave us a deterministic interpretation of quantum mechanics which as a consequence,

---

[87] "Probability and Uncertainty — the Quantum Mechanical View of Nature", video lecture by Richard Feynman given at Cornell University, 1964

implies something even more bizarre, that we live in a parallel universe.

38. Hugh Everett (1930-1982)

"for the many worlds interpretation of quantum mechanics"

That whenever you make a decision, another parallel universe exists in which you might have taken a different decision, sounds like the plot of a science fiction story. But it is an interpretation of quantum mechanics that was developed by Hugh Everett for his PhD thesis. Ridiculed at the time, it is now considered to be one of the leading interpretations of quantum mechanics.

Hugh Everett was born in Washington D.C. At the age of 12 he wrote letters to Albert Einstein raising the question of whether it was something random or unifying that held the universe together. Einstein responded in a 1943 letter *"Dear Hugh: There is no such thing like an irresistible force and immovable body. But there seems to be a very stubborn boy who has forced his way victoriously through strange difficulties created by himself for this purpose. Sincerely yours, A. Einstein."* After obtaining a degree in chemical engineering and performing well in mathematics, in 1953 Everett received a fellowship for his PhD studies at Princeton University under the supervision of the eminent physicist John Wheeler. Several of Wheeler's PhD students, such as Richard Feynman and Kip Thorne, would go on to win a Nobel Prize. I believe that Everett deserved that honour too.

The standard interpretation of quantum mechanics is called the Copenhagen interpretation, just because Niels Bohr and his assistant Werner Heisenberg were in Copenhagen. Bohr and Heisenberg received Nobel Prizes in 1922 and 1932 respectively, for their contributions to the theory of quantum mechanics. They also interpreted the theory in a very particular way. Bohr advocated that matter exists in a mixture of all possible quantum states until it is observed. At that point the so-called wave function of Schrödinger, is said to collapse and picks a state according to a random probability. This duality is named superposition. I dislike the idea of nature having a built-in random number generator as much as Einstein illustrated in his famous quote, often paraphrased as 'god does not play dice'[88].

In the Copenhagen interpretation of the double slit experiment described in the last chapter, it is asserted that the electron has an inherent probability wave that it carries with it. The wave and particle nature of objects can be regarded as complementary aspects of a single reality, like the two sides of a coin. An electron, for example, can behave sometimes as a wave and sometimes as a particle, but never as both together. It behaves as a probability wave until it is observed, or measured. Until it is measured, or observed, it has no reality. The probability wave that describes the electron, passes through both slits and it interferes with itself. In essence, the particle's journey is a superposition of all possible trajectories. The particle simultaneously takes all possible paths and passes through both slits. Those histories converge on one possible outcome when possibility space collapses to a single reality upon measurements. It is a non-deterministic universe since there is no way of predicting exactly where any given electron will land. The superposition of the histories of the particle's

---

[88] In 1926 Einstein wrote a letter to Max Born: "Die Quantenmechanik ist sehr achtung-gebietend. Aber eine innere Stimme sagt mir, daß das doch nicht der wahre Jakob ist. Die Theorie liefert viel, aber dem Geheimnis des Alten bringt sie uns kaum näher. Jedenfalls bin ich überzeugt, daß der nicht würfelt." This is often translated into English as "Quantum mechanics is certainly imposing. But an inner voice tells me that it is not yet the real thing. The theory says a lot, but does not really bring us any closer to the secret of the "old one." I, at any rate, am convinced that He does not throw dice."

trajectory merge into the single timeline of the observer's reality. Its position only becomes definite when it hits the detector. At that point the wave function is said to collapse and it becomes localised as a particle.

Schrödinger derived the famous 'wave function' that describes basically everything in the universe. He did not like the idea of his wave function collapsing – there was nothing in his theory about such a collapse - and he gave the following thought experiment to illustrate its philosophical problems. He proposed a scenario with a cat in a locked steel chamber, wherein the cat's life or death depends on the state of a radioactive atom, on whether or not that atom had decayed and emitted radiation. The random decay of that atom could be set up to unlock a vial of cyanide that would kill anything in the box. Being in a system together closed from the outside world, the cat and the radioactive atom are in a quantum superposition of states. The radioactive decay is a purely quantum process, and until it is observed it exists in a superposition of states in which it both has and hasn't decayed. The cat is also in superposition with the particle that undergoes radioactive decay. The Copenhagen interpretation implies that the cat remains both alive and dead until the box is opened and its state has been observed. Here, 'the observer', is anything that interacts with the contents of the closed box, such as a photon of light or a human observer.

This leaves us with some obvious questions. How can the cat be both alive and dead if you don't look at it? What actually collapses and why do the other states just disappear? Why should measurement or observation determine reality? And why do quantum thought experiments involve suffering cats?! (Schrödinger's daughter is often quoted as saying *"I think my father just didn't like cats."*) Hugh Everett tackled this problem for his PhD thesis and came up with a radical suggestion, one that I think Schrödinger himself would have approved. In Dublin in 1952, Erwin Schrödinger gave a lecture in which he jokingly warned his audience that what he was about to say might *"seem lunatic"*. He said that when his equations seemed to describe several different histories, these were not alternatives, but all really happen simultaneously. But he never took the idea further. Independently, and bravely, Hugh Everett developed an interpretation of quantum mechanics that described the reality of alternative paths through time.

Everett completed his PhD thesis in 1957 titled 'On the foundations of quantum mechanics' which he published as a single short paper 'Relative state' formulation of quantum mechanics'. The editor of the journal in which he published in, Reviews of Modern Physics, was the well-known physicist Bryce DeWitt, who regarded the work positively. Everett had previously written a longer text titled 'Wave mechanics without probability' that would only be published much later. Wheeler had told Everett that his lengthy text was simply too controversial to publish.

Everett thought the simplest interpretation of quantum mechanics was to just follow the math. He asserted that the wave function is real and never collapses causing the other possible quantum states to mysteriously vanish. He simply assumed that the wave function represents reality entirely and exactly and always obeys the form written down by Schrödinger. In Everett's description of quantum mechanics, Schrodinger's cat is alive in one universe and dead in a parallel universe. The moment you open the box and look inside, the universe branches into two paths. In one universe you are looking at a healthy living cat, in another you sadly see a dead cat.

In Everett's interpretation the universe is deterministic and entirely determined by the wave function. It implies that we live in a near infinite, or infinite number of universes, all superimposed in the same physical space but isolated from each other and evolving independently.

Wheeler sent a draft of Everett's work to Neils Bohr in Copenhagen and went to visit him in person to discuss the ideas. Wheeler reported that Bohr was not very receptive of the ideas of Everett. Discouraged, Everett applied for a job at the Pentagon. In the meantime, Wheeler wrote a letter to Everett encouraging him to visit Bohr: *"Unless and until you have fought out the issues of interpretation one by one with Bohr, I won't feel happy about the conclusions to be drawn from a piece of work as far reaching as yours"*. He added *"So in a way your thesis is all done; in another way, the hardest part of the work is just beginning."*[89]

After starting work at the Pentagon, Everett did visit Bohr for six weeks in 1959, about which Everett later stated *"that was a hell...doomed from the beginning"*[90]. Bohr's colleague in Copenhagen, Léon Rosenfeld, later described Everett's visit in a 1972 letter to the Dutch physicist Frederik Belinfante: *"With regard to Everett neither I nor even Niels Bohr could have any patience with him, when he visited us in Copenhagen more than 12 years ago in order to sell the hopelessly wrong ideas he had been encouraged, most unwisely, by Wheeler to develop. He was*

---

[89] "The origin of the Everettian heresy", 2009 Stefano Osnaghi et.al. Studies in the history and philosophy of science, 40, 2, 97-123.

[90] Hugh Everett interviewed by Charles Misner, 1977, Oral History Interviews, American Institute of Physics

indescribably stupid and could not understand the simplest things in quantum mechanics."[91]

Everett's ideas were forgotten for over a decade. At the pentagon's weapons system evaluation group, much of his work was classified, but he did publish an influential study together with George Pugh titled 'The Distribution and Effects of Fallout in Large Nuclear Weapon Campaigns'. This was mentioned in the Nobel Prize speech of Linus Pauling. Everett and Pugh estimated that 90% of Americans would be dead sixty days after a large-scale nuclear war. Pauling states *"The fate of the living is suggested by the following statement by Everett and Pugh: "Finally, it must be pointed out that the total casualties at sixty days may not be indicative of the ultimate casualties. Such delayed effects as the disorganization of society, disruption of communications, extinction of livestock, genetic damage, and the slow development of radiation poisoning from the ingestion of radioactive materials may significantly increase the ultimate toll."*[92] After working at the pentagon Everett founded several companies to apply military modelling solutions to civilian problems.

Everett's work was brought to the attention of the world in 1973 when DeWitt published a book 'The many worlds interpretation of quantum mechanics', which containing Everett's original and longer unpublished thesis text. DeWitt succinctly wrote *"every quantum transition taking place in every star, in every galaxy, in every remote corner of the universe is splitting your local world on Earth into myriads of copies of*

---

[91] "The origin of the Everettian heresy", 2009 Stefano Osnaghi et.al. Studies in the history and philosophy of science, 40, 2, 97-123.

[92] Linus Pauling, Nobel Lecture, December 11, 1963

itself."[93] The book sold out and the idea of many worlds, or parallel universes, began to appear in science fiction stories. In 1977, Wheeler invited Everett to talk about his results at a conference, which would be the first time that Everett would talk about his research to a large audience. Everett obliged and Wheeler's student David Deutsch was present and found the ideas of Everett inspiring. He extended the theory and popularised the ideas further.

Wheeler encouraged Everett to return to a physics career but Everett wanted nothing more to do with science. Later Wheeler, who had initially supported Everett's ideas, backtracked and in 1979 stated that he had reluctantly given up his support of the theory as it created too great a load of metaphysical baggage.

Today, Everett's many worlds interpretation is hotly debated amongst philosophers, quantum physicists and the public alike. His one publication on the topic has now received over 4000 citations by other scientific works. At a 1997 conference of theoretical physicists, a poll was held on which of the many interpretations of quantum mechanics they preferred. The Copenhagen interpretation came first followed closely by Everett's many worlds. Many physicists think that the many worlds interpretation is just too bizarre. Others, such as string theorist Brian Green, theoretical physicist Sean Carroll and the late Stephen Hawking, think that it is the only plausible interpretation of quantum mechanics.

Some scientists and philosophers argue that the many worlds theory is worthless because it can't be tested. That a scientific theory should be testable, was advocated by the

---

[93] 'The Many Worlds Interpretation of Quantum Mechanics', 1973 edited by Bryce DeWitt and Neill Graham, Princeton University Press, page 161

famous British philosopher Karl Popper. However, Popper himself was a vocal critique of the Copenhagen interpretation for six decades. In his 1982 text 'Quantum Theory and the Schism in Physics', Popper endorsed Everett's interpretation and came up with ideas on how to test the reality of the Copenhagen interpretation. And there are some ideas as to how the existence of parallel universes could be tested. David Deutsch argues that an intelligent quantum computer would be able to remember the experience of temporarily existing in parallel realities.

In his published paper, Everett wrote that his formulation of quantum mechanics could prove a fruitful framework for the quantization of general relativity. Some cosmologists have taken his work a step further in which they describe the entire universe by a single wave function which describes all possible universes at all possible times. In such a universe, all physical possibilities simultaneously exist as snapshots frozen in time. Time itself doesn't flow, rather past and future events at all instances in time, just exist.

Everett died of a heart attack in 1982 at the age of 51. It is thought that his lifestyle of chain smoking and alcohol excess led to his early death. He asked for his ashes to be thrown out in the trash after his death, a wish kept by his wife. His daughter Elizabeth committed suicide in 1996. In her suicide note she wrote that she wanted to end up in the same parallel universe as her daddy. Everett's son, Mark, is the main singer and songwriter for the band Eels. His famous album 'Electro-Shock Blues' was written during this period, with song titles such as 'Elizabeth on the bathroom floor' and 'Going to your funeral'. In a 2007 BBC documentary, 'Parallel Worlds, Parallel Lives', Mark mentions how he wasn't even aware of his father's status as a brilliant physicist until after his death.

39. Stephen Hawking (1942-2018)

"for showing that even black holes will not last forever"

In the far future all that is left of our galaxy is its central supermassive black hole. Will that exist alone for an eternity of time in the darkness our universe will become? After all, nothing can escape from a black hole we are told. But that is not the case, energy slowly leaks from black holes such that even they will vanish in the far future. That black holes slowly evaporate was one of the most important results discovered by the English theoretical physicist Stephen Hawking. Hawking proved that indeed, nothing lasts forever.

I first met Stephen Hawking whilst studying physics at Newcastle University. I was in the library with a friend and we heard this strange robotic voice coming closer from down the corridor. *"What on Earth is that?"* we both exclaimed. The strange robotic voice answered *"Stephen Hawking"* as he wheeled into our room leaving us very embarrassed. It was 1987 and Hawking was visiting Newcastle University to receive an honorary degree. Two years before, Hawking had needed a life-saving tracheostomy that left him unable to speak. He was very proud of his computer aided speech generator that allowed him to communicate and which became one of his recognisable traits, but we were unaware of this at the time. Hawking was already famous amongst scientists for his work on singularities and black holes, but he was still happy to take the time to talk to us excited young undergraduates. The following year Hawking's fame exploded after his 1988 book 'A brief history of time' became a bestseller and sold over 25 million copies. It is sometimes called the most popular book never read because of its terse complexity. A decade later I met Hawking several times as we collaborated on a research computer to simulate the universe at the University of Cambridge. During one dinner I asked him if he remembered talking to some rude students at

Newcastle University. There was a lengthy silence, but I was relieved when his computer voice said "no".

Stephen Hawking was born on January 8th 1942, the same day that Galileo died 300 years before. During his first years at school Hawking was rarely above average in the class, although his classmates gave him the nickname 'Einstein'. At high school it was his math teacher that inspired Hawking to study mathematics and physics. He won a scholarship to study physics at Oxford University, where, he stated, he managed to get his degree with little work and studied just an hour a day because he found the work ridiculously easy. Towards the end of his university studies, he noticed he was getting increasingly clumsy and during his PhD at Cambridge he was diagnosed with motor-neuron disease and was told he would only have a few years to live.

It was his meeting with his first wife that motivated him and gave him something to live for. He decided to start working hard for the first time in his life and was surprised to find that he enjoyed it. Although he was initially disappointed upon joining Cambridge that his PhD supervisor was not Fred Hoyle, he would soon challenge Hoyle's alternative theory of gravitation with his first publication aged 23. In Hoyle's steady state cosmological model, the universe is eternally expanding with new matter constantly being created. Hawking showed mathematically that this theory was self-contradictory.

Hawking then investigated far reaching problems in cosmology that are mathematically complex: "Does the universe have a beginning and did an initial singularity exist?" The presence of a singularity in the equations that described the big bang was noted by Lemaître in 1930. That the universe would be infinitely dense and hot at the time $t=0$

was a feature disliked by many scientists. In his PhD thesis Hawking showed that singularities are unavoidable in such models – that time itself had to start at the big bang.

Hawking pushed our knowledge of mathematics and physics to its limits to show where it failed. He demonstrated the need for a unified description of general relativity and quantum mechanics. His theoretical work on black holes led to fundamental laws and rules that describe their behaviour in time and space. His aim was to understand the entire universe, its past and future, with science. When he lost the use of his hands in the mid-1970s he developed an amazing ability to visualise and manipulate complex equations and geometric constructs in his head and much of his greatest research ensued.

His most well-known research published in 1974 involved connecting the quantum world, that describes particles and atoms, with the curved space-time surrounding black holes that is described by general relativity. Hawking himself was surprised by what the equations yielded. Black holes are not black timeless entities. Rather, they radiate energy and as a consequence they slowly lose mass until they completely disappear in a final intense burst of light.

In the far future, around 10^{22} years (ten thousand billion billion years) from now, our galaxy of dead stars, failed stars and cold planets has evaporated – gravitational encounters between these objects either eject them from the gravitational confines of the galaxy or send them into the central supermassive black hole, the final remaining object in our galaxy. In 1971 Roger Penrose showed how energy could be extracted from such an object which could provide enough fuel for a highly advanced civilisation to exist for an enormous length of time. Just a few years later Hawking showed that the lifetime of a black hole depends on the cube of the mass of the black hole. If our Sun collapsed into a black hole, it would radiate away its mass on a timescale of about 10^{61} years. If by the end of our galaxy its central black hole may have grown by accreting dead stars and planets to a mass over 10^{10} times that of our Sun, it will ultimately evaporate away in about 10^{100} years. And that's it. The last physical object from which we could obtain energy in the far future is gone.

How can a black hole, that we learn curves spacetime so intensely that even light cannot escape, lose mass? Hawking described this process in relatively simple terms. The space around us, or around a black hole, is not empty. As we learned in the chapter on Tryon, the vacuum of space is filled with energy fields and that gives rise to a sea of activity. Virtual particles and their virtual anti-particles appear and disappear on timescales so short that they do not violate conservation laws. But if this happens close to the edge of the black hole, one virtual particle may enter the black hole and one may escape. If one particle falls into the black hole and one particle escapes, it seems that the mass of the black hole should not decrease. Hawking explains this by stating that the escaping particle has a positive energy whilst the captured particle has negative energy. Ultimately the black hole loses mass and evaporates away to nothing.

For decades after this picture has been used to describe how black holes evolve in time. But it is just an analogy and it is quite a poor one. Virtual particles are not real entities, but mathematical tools used to calculate the behaviour of quantum fields. And most of the predicted mass loss of the black hole arises from photons and not particles like electrons. Moreover, most of the radiated emission occurs at large distances from the black hole and at very low energies. The physical interpretation of how black holes lose mass is still debated amongst theoretical physicists. They all agree that this is what the equations show and what must happen, but they disagree as to how it happens. One interpretation states that this escape of matter can happen through quantum tunnelling – that strange fact of quantum mechanics which states you can't exactly pin down where a particle is and it may suddenly find itself in a region where classical physics says it should not be.

Despite Hawking's physical condition he had a positive outlook to life. He made the best use of his fame, from a guest appearance on Star Trek to making a zero-gravity flight. In 1979 he became the 17th professor to hold the Lucasian chair in Cambridge, over three centuries after Isaac Newton had become its second holder. He was widely recognised for his research and received many awards and prizes, from the three-million-dollar breakthrough prize to the individual Einstein, Eddington and Dirac medals. Yet despite his many contributions to science and our understanding of the universe, many people are surprised to learn that he never received a Nobel Prize for his work.

Hawking radiation has such a low energy and is such a slow process that it is unlikely to be confirmed for several centuries until we can get up close to a black hole. Perhaps for this reason Hawking was always an unlikely candidate for the Nobel Prize. As I mentioned in the foreword, the Committee require proof or evidence of a theory, at least in most cases.

Hawking died peacefully aged 76. His ashes lie in Westminster Abbey, in between the graves of Isaac Newton and Charles Darwin. His memorial stone is inscribed with the words *"Here lies what was mortal of Stephen Hawking 1942–2018"*. It also contains the equation he derived for the temperature of a black hole. Hawking had requested that his famous equation for the entropy of a black hole should be inscribed, $S=kc^3A/4\hbar G$ (similar to Boltzmann's gravestone which hosts his formula for entropy $S=k \log W$). The correct inscription was made on his memorial in Cambridge's Gonville & Caius College along with Hawking's own words *"Remember to look up at the stars and not down at your feet"*, his full quote continues *"Try to make sense of what you see and wonder about what makes the universe exist. Be curious. And however difficult life may seem, there is always something you can do and succeed at. It matters that you don't just give up"*[94].

---

[94] 'Brief answers to big questions', 2018, Stephen Hawking, pub. Hodder & Stoughton, page 157.

40. Freeman Dyson (1923-2020)

"for contributions to quantum field theory and for his visionary work on the future of life in our universe"

Quantum field theory was developed in the 1950s and it is the deepest level of understanding of the workings of nature that we have. It was mainly developed by four theoretical physicists, and of these only the British physicist, Freeman Dyson was not awarded a Nobel Prize.

At the age of four Freeman Dyson tried to calculate the number of atoms inside our Sun; when he was eight he wrote a science fiction story about a journey to the Moon. These interests would persist throughout his scientific career. At boarding school he wanted to read Ivan Vinogradov's 'Introduction to the theory of numbers' but it was only available in Russian. Aged fifteen he learned Russian on his own and translated the text into English.

Dyson won a scholarship to attend Cambridge University but his studies were interrupted by World War II. He was assigned to applying his mathematical skills to analyse the effectiveness of different military strategies for the Royal Air Force. This experience gave him a lifelong commitment to ensuring that such a war never occurred again. Following the war, he returned to Trinity College and earned a degree in mathematics. He stayed at Cambridge for three more years and wrote several papers on number theory. He was then awarded a Commonwealth Fellowship to work with Hans Bethe at Cornell University, the place where many of the most brilliant Los Alamos scientists went to following the war. It is there he met Richard Feynman.

In 1948 the theoretical physicists Julian Schwinger and Richard Feynman were advocating two different approaches to understanding the nature of particles. The previous year the experimental physicist Willis Lamb had made precise measurements of the energy levels of a hydrogen atom and the existing quantum theory could not explain the results. It turns out that you have to consider the interaction between vacuum energy fluctuations and the hydrogens electron. Lamb would receive the 1955 Nobel Prize in physics for his experiments that led to the theories of quantum electrodynamics and quantum field theory.

Feynman proposed a simple set of diagrams that described the interactions of particles as they moved through space and time. Schwinger developed a complex mathematical formalism that few scientists understood. A third approach was developed independently at the same time by Japanese physicist Sin-Itiro Tomonaga who developed a mathematical way of dealing with the infinities that keep arising in such theories. Dyson realised how to reconcile the three approaches – he translated Feynman's diagrams into mathematics and showed that they captured the essence of Schwinger's abstract algebra, which itself was equivalent to Tomonaga's mathematical tricks. It was the birth of quantum electrodynamics and modern particle physics. Quantum electrodynamics is the relativistic quantum field theory of electrodynamics that describes how light and matter interact. It is a subset of quantum field theory that I described in the chapter on Tryon.

In 1965 Nobel Prizes were duly awarded to Feynman, Schwinger and Tomonaga, but perhaps because of the rule of three recipients, Dyson missed out. Later, Dyson would say that *"it is better to be asked why you didn't get a Nobel Prize than why you did"!*[95]

Still without a PhD, Dyson was offered a professorship at Columbia, but he had to return to the UK according to the rules of the Commonwealth Fellowship. In 1951 he returned to

---

[95] Freeman Dyson, 2010 video interview by Austin Allen for Big Think

Cornell as a professor, despite never completing his PhD, and two years later he was offered a lifelong position at IAS Princeton where he stayed for six more decades. He was happy to enjoy the freedom of being 'a permanent student' and wrote that he disliked the entire PhD system. Dyson thought that the PhD forces students to waste years of the lives to obtain a meaningless piece of paper that states they are a scientist, and that it tragically discourages women from become scientists. And I think that is a valid statement – it is a relic of the 19th century German university system only useful for the small subject of students that actually go on to become professors.

Dyson was one of those scientists who, like Gamow and Zwicky, moved from one problem to the next, even if it were in a different domain of research. However, Dyson had the mathematical skills to take a problem and work out its complete solution. The only requirements for Dyson were that the problem needed to be interesting and simply stated. For example, he became interested in the peaceful uses of nuclear energy and played a key role in the design of a self-regulating nuclear reactor that would not explode. That design is now one of the most widely used nuclear research reactors in the world.

Some problems seem simple but turn out to be incredibly hard to prove. One of the most complex calculations Dyson performed, in collaboration with Andrew Lenard, was to prove that solid objects can exist. They showed that solid objects should collapse to an incredibly dense state if you only consider the attractive and repulsive forces between the electron and the nuclei. They went on to prove that it is the Pauli exclusion principle that enables objects to maintain their structure. The Pauli exclusion principle was formulated by the

Austrian scientist Wolfgang Pauli in 1925, who showed that identical electrons cannot occupy the same physical space at the same time. It is the reason why objects are solid to our touch, despite atoms consisting of mostly empty space.

Disappointed with NASA's ambition of only visiting the Moon, Dyson worked on Project Orion at Los Alamos National Laboratory. It was a mission to design a spacecraft that could carry people to the distant planets and the stars beyond. The Orion spacecraft was to be powered by a series of nuclear explosions behind a specially designed spacecraft. A huge steel or lead shock-absorbing plate would take the impact of the bomb debris, which occurs in a few nanoseconds, spreading the impulse over a few seconds. The spacecraft would literally be pushed forward by the blast wave of particles coming from the nuclear explosion. After about a thousand blasts, the spacecraft would be accelerated up to ten percent of the speed of light. Allowing for speedup and slowdown, the journey to the next star could take as little as 50 years; it could thus be made in a single human lifetime. Dyson designed missions of several sizes, up to the Super Orion craft, a 400-metre vessel that could hold a small city. However, interest in nuclear pulse propulsion declined in 1963 following the Nuclear Test Ban Treaty signed by the United States, the Soviet Union and the United Kingdom. This agreement banned nuclear weapons tests in the atmosphere, underwater and in outer space.

Dyson thought about life in the context of the universe throughout his career. In 1959 Giuseppe Cocconi and Philip Morrison realised that the radio telescopes used by astronomers were powerful enough to detect artificial radio signals transmitted by alien civilisations on other worlds. This inspired Frank Drake to begin the SETI project – the search for extra-terrestrial intelligence, and to write down his famous 'Drake equation' which estimates the number of potential civilisations with whom we could radio communicate.

Dyson wondered whether or not it would be possible to detect advanced civilisations that did not use radio communications. He argued that any civilisation that used a large amount of energy would be unable to contain waste heat and that this would be detectable. In 1960 he published a short paper in the prestigious journal Science titled 'Search for Artificial Stellar Sources of Infrared Radiation'. Dyson describes a way to detect the presence of highly advanced alien civilisations. He envisaged that the energy needs of a civilisation will inevitably grow larger and larger. Eventually more energy would be needed than can be harnessed from the starlight which illuminates their planet. Sooner or later, an advanced alien civilisation will need to capture more of their star's energy. To accomplish this, they would surround their star with giant artificial structures. Such constructions that may partially or entirely surround stars to gather their energy are now known as 'Dyson Spheres'. But Dyson himself acknowledges that the idea was taken from the 1937 book 'Star Maker' written by the science fiction writer Olaf Stapledon and suggested they should be called 'Stapledon Spheres'.

Dyson realised that any structures that collect the energy of a star will become warm and will radiate waste heat. He calculated that they should be easy to detect with infra-red telescopes. But since water vapour in our atmosphere absorbs light at these wavelengths, the search for Dyson Spheres had to wait until telescopes were placed on high mountain tops or in space. Any Dyson Spheres that completely enshrouded their stars would have been visible to a distance of 1000 light years from the Sun. Within that distance there are a million stars like our Sun. The IRAS and WISE space telescopes mapped the entire sky in the infra-red frequencies. So far, no glowing Dyson Spheres have been discovered.

In 1979 Dyson pondered the ultimate fate of life in a universe that expands forever, one of the first serious eschatological studies. When all the stars have died, and all the black holes have evaporated, what happens to life? To have a conscious thought needs energy and there are no more energy sources to be found. As time ticks on into the future, it becomes more and more difficult for life to function – even if life in the far future resembles a computational machine. Dyson devised a clever way in which an intelligent being could think an infinite number of thoughts in an open universe using a finite amount of energy. The intelligent beings would begin by storing a certain amount of energy, and they would then use some fraction, say one- half, of this energy to power their thoughts. When the energy gradient created by unleashing this fraction of the stored fuel was exhausted, the beings would enter a state of zero-energy-consumption hibernation as the universe cooled around them. Once the universe had cooled sufficiently, an alarm clock would wake up the life. Half of the remaining half (one quarter of the original energy) of the intelligent beings' fuel reserves would once again be released, powering a brief period of thought once more. This would continue, with smaller and smaller amounts of energy being released. The time between thoughts would become longer and longer, but there would still be an infinite number of them. The only problem I see with this idea is that any alarm clock is ultimately destined to fail due to quantum mechanical fluctuations that randomly rearranges its components.

Amongst his many awards for his scientific work, Dyson also received the Templeton Prize for Progress in Religion in 2000, for his writings on the relationship between religion and science. Several of my colleagues, including Richard Dawkins, criticised his acceptance of this award, which has a monetary value larger than the Nobel Prize. They argue that it is used to blur the well-defined boundary between science and religion. Dyson called himself a christian but was rather agnostic in his opposition of traditional views of a god, writing *"I have no use for a theology that claims to know the answers to deep questions but bases its arguments on the beliefs of a single tribe. I am a practicing Christian but not a believing Christian. To me, to worship God means to recognize that mind and intelligence are woven into the fabric of our universe in a way that altogether surpasses our comprehension."*[96]

In 2006 Dyson published a collection of his essays in his book 'The scientist as a rebel'. There he wrote *"We should try to introduce our children to science today as a rebellion against poverty and ugliness and militarism and economic injustice"*[97]. Dyson retired from Princeton's Institute for Advanced Study in 1994. He remained active in his thoughts and ideas until his death aged 96 in 2020.

---

[96] Science & Religion: No Ends in Sight - a review by Freeman Dyson of The God of Hope and the End of the World by John Polkinghorne". The New York Review of Books. 28 March 2002.

[97] The Scientist as a Rebel' 2006, Freeman Dyson, pub. New York Review Books, page 7.

41. Erast Gliner (1923-2021)

"for the theory of inflation"

That before the hot big bang, our entire universe was empty of matter and rapidly expanded from the volume of a subatomic particle to the size of a football; this is called inflation. The theory of an inflationary universe was developed from 1965-1983 and is considered by many cosmologists to be the greatest step forward in our understanding of the early universe since the work of Lemaître and Gamow. The theory was designed to solve the initial singularity problem, to explain the geometry of spacetime and why the universe looks the same in all directions. It also led to a way to generate irregularities in the matter distribution that ultimately led to stars and life, as well as the concept of the multiverse. Inflation is now considered to be part of the big bang model for the universe. However, some scientists disagree, stating that it does not solve any problems and should not be treated as a theory because it leads to predictions that can't be tested.

What is the theory of inflation, who came up with the idea and why is it important? These are all difficult questions to explain succinctly but let me try.

The theory of inflation is most widely credited to the American scientists Alan Guth and Andrei Linde who received the 2004 one-million-dollar Gruber prize for this theory, as well as the three-million-dollar breakthrough prize in 2012. Not an insignificant little extra income for a scientist! However, the full story is more complex, involves many other scientists and begins with the Soviet renaissance in cosmology. Of all the scientists who developed the theory of inflation, one in particular stands out – the first to propose such a theory and who has been widely ignored – Erast Gliner.

In the Soviet Union the idea of a universe with a beginning did not fit well with the communist ideologies. Dialectical materialism was the imposed philosophy of science based on the writings of Marx, Engels and Lenin. That the universe had a beginning implied a cosmic creation which was taken to be religious. Their philosophy implied that the universe should be eternal and infinite – the sanctioned view of materialism. In the People's Republic of China cosmology became a forbidden science in Mao Zedong's empire that had its own version of dialectical materialism. Relativistic cosmology was declared reactionary and an anti-socialist pseudo-science. Soviet scientists simply did not work on developing the big bang theory further, despite scientists such as Alexander Friedman's earlier contributions to the theory. It was only in the 1960s after the death of Stalin that Soviet researchers begin to explore cosmological models again.

Erast Gliner was born in Kiev in 1923, which today is in Ukraine. He began to study chemistry at Leningrad University but that was interrupted by World War II. Over a million civilians were mobilised in constructing fortifications of the city, including Gliner who like many, suffered physically from the effort. He was excused from serving in the military due to his condition - he had developed severe dystrophia, but he volunteered for the army anyway. He was wounded twice and received two medals of honour. After returning to action he was wounded a third time and lost his right arm. He returned to Leningrad University and began to study physics. In 1945 he was arrested for 'participation in the activity of an anti-Soviet group', which was no more than a student literary circle. Gliner was sent to the infamous Soviet Gulag – forced labour camps which held millions of political prisoners, intellectual dissidents and anyone who the KGB chose to send there. Gliner spent ten long years in the Gulag. Following the death of Stalin, the Gulags were slowly disbanded and Gliner returned to his studies in 1955.

Whilst at Leningrad in the 1960s, Gliner wrote several papers on relativistic physics and cosmology. He submitted his works as a thesis, but to qualify for the Russian doctorate he needed three members of the Academy to approve the work. Despite strong support from two scientists, one of whom was the theoretical physicist Andrei Sakharov who would later obtain the Nobel Peace Prize, the most important scientist in Russia was Yakov Zeldovich and he refused to support his thesis – it was too revolutionary[98]. Gliner's colleagues secretly arranged for his thesis documents to be taken to Tartu University in Estonia where he was awarded his doctorate in 1972. Later, in the 1980s when the theory of inflation was reborn, Zeldovich admitted his mistake. So, what did Gliner's thesis contain?

At this time it was widely known that the equations developed by the likes of Friedmann and Lemaitre to describe our universe were known to have a problem – if you keep extrapolating them backwards you reach a singularity at time t=0. The singularity, an infinity, arises because the size of the universe becomes zero and its energy density becomes infinite. We do not believe this was the state of the universe, rather it indicates that we do not understand what was going on during the first moments of the universe. The singularity was a problem but Gliner came up with a possible solution.

Gliner realised that extrapolating backwards in time the universe would reach densities such that it could be described by a relativistic vacuum. He brought back Einstein's cosmological constant and associated it with the energy

---

[98] A. Silbergleit and A.D. Chernin, "Why does the Universe Expand? A tribute to E.B. Gliner", Springer, 2017.

density of the vacuum. Einstein's constant acts like a repulsive gravitational force and could cause a universe empty of matter to undergo an incredibly fast and potentially huge exponential expansion. Gliner argued that our entire universe began as a miniscule piece of spacetime empty of matter, a tiny fraction of the size of an atom, which then rapidly expanded faster than the speed of light![99]

At some point, Gliner speculated that the universe would transition from an 'empty' vacuum energy dominated universe to the phase we call the big bang. At this point the energy in the vacuum is converted into matter and radiation. The dynamics of a universe empty of matter had previously been solved by the Dutch physicist Willem de Sitter. During this rapid expansion the vacuum energy density remains constant, no matter how far back in time you go. In such a universe there would be no infinite density of matter, no singularity and no origin in time[100].

Gliner published his results in a series of papers from 1965 to 1975, some in collaboration with his student Irina Dymnikova. He left Russia in 1980 and spent some time working at the University of Colorado, Boulder McDonnell and the Center for Space Science at Washington University St. Louis.

According to theoretical physicist and Nobel Laureate, Vitaly Ginzburg, Gliner was a successful researcher until he

---

[99] The expansion of space occurs far faster than the speed of light, but that's ok since it is only space that is expanding. That physical objects are not accelerated beyond the speed of light, or that information passes faster than light, is the key requirement of Einstein's theory of relativity.

[100] If you extrapolate an exponential expansion back in time, you never reach zero volume!

apparently gave a seminar on inflation in 1987 at an unspecified world-famous university in America. Ginzburg was the chief editor of the English version of the Russian journal of Physics. In a foreword to one of Gliner's papers published in 2002 he writes[101]: *"His ideas, close to those in the paper below, apparently turned out to be objectionable to some prosperous cosmology authorities in the US. Alas, this cost Gliner his job, and his pension too, because his dismissal left him one year short of the necessary period of service. [...] as things stand, however, E.B. Gliner has no affiliations, cannot attend scientific conferences, and is denied access to the university online services."* From 1987 Gliner lived in San Francisco, unrecognised for his contributions to early universe cosmology. He passed away in 2021 aged 98.

Gliner called the rapid expansion of the early universe a 'blow-up' phase. Later, this would be named 'inflation' by Alan Guth. Gliner didn't appreciate that in such a model he could have solved several additional problems in cosmology that we will soon hear about. Gliner's model was also a bit ad-hoc, primarily because the transition from an inflationary phase into the standard matter-radiation dominated big bang was just presumed without a physical basis.

Alexei Starobinsky was a student of Yakov Zeldovich and he took the ideas of Gliner further. In 1978 he began to consider how developments in the emerging theory of quantum gravity would affect the early universe. Starobinsky found that the early universe could have begun exponentially expanding – a so called De Sitter phase, before undergoing oscillations that caused a transition into an expanding universe filled with matter and light. He also realised that this

---

[101] "Inflationary universe and the vacuumlike state of physical medium", 2002, Physics – Uspekhi 45, 213-220.

early exponential expansion would leave an observational signature – gravitational waves with a specific set of frequencies. These would be difficult to detect but could prove to be an important test of the theory.

Inspired by the work of Starobinsky, in 1981 the Russian cosmologists Viatcheslav Mukhanov and Gennady Chibisov published a study showing how a rapidly expanding phase in our early universe could give rise to the fluctuations observed in the cosmic microwave background. The quantum field that was causing space to inflate, would like any quantum field, undergo fluctuations. Space would be rippling with tiny variations in energy on the quantum scale. When space inflated, any tiny variations in energy would also become amplified. At the end of inflation, the quantum fluctuations would be amplified into much larger irregularities in the energy field, which later would become irregularities in the matter distribution. Quantum fluctuations in the early universe would be expanded from microscopic to macroscopic scales due to the incredible expansion of space. These fluctuations are what we see in the microwave background that I described earlier, and they ultimately evolved into galaxies, stars, planets and life. What a beautiful idea, that everything we see in our universe began as a quantum fluctuation!

One of the key results from the Hubble Space Telescope were the observations that revealed that the universe has evolved over time – that these initial quantum fluctuations have grown thanks to gravity. By looking at the most distant reaches of the universe you find younger, less evolved and smaller galaxies that are forming far more stars than today. By taking exposures of the same patch of the night sky for over a million seconds, the Hubble Space Telescope has detected tiny galaxies that were forming as long as 13.4 billion years ago – a time when the universe was just three percent of its present age. The same observations have revealed that the rate at which stars formed peaked around 10 billion years ago and has since declined. Most of the stars that could form have already formed long ago. This observed development of cosmic structure over time is the fourth pillar of evidence for the big bang.

The theory of inflation was again proposed in the early 1980s simultaneously and independently by Demosthenes Kazanas, Alan Guth, and Katsuhiko Sato. They speculated that inflation occurred around 10^{-34} seconds after the appearance of our universe, when three of the fundamental forces of nature (electromagnetism and the weak and strong forces) acted as a single symmetric unified force field. As the universe expanded the energy of the field dropped until it reached a value at which it spontaneously split into several new fields, a so-called 'symmetry breaking phase transition' that resulted in separate distinct forces. The quantum field that drove inflation was thought to be the same field that gave rise to the Higgs particle. Most scientists associate Guth with being the inventor of the modern theory of inflation. Kazanas was actually the first to publish in 1980, although it is clear that these authors developed their ideas independently. And none of them referenced the work of Gliner or Starobinsky. However, Kazanas and Guth realised that an inflationary early universe would solve two major puzzles – the so called 'horizon problem' and the 'flatness problem'.

The theory of inflation provides a mechanism to explain the cosmological principle – why the universe appears the same in all places and directions. This is known as the horizon problem. Without inflation, two widely separated regions in the observable universe would never have been close enough to feel each other's presence. Gravity and light would simply not have had time to propagate between the two. And during the early universe, two separated regions may end up with different amounts of energy and matter. However, the cosmic microwave background shows that the temperature of the universe was the same everywhere to a precision of one part in 100,000. Inflation solves this problem since the region that gave rise to our entire visible universe, was once a tiny region that was 'causally connected' – a tiny region small enough that information had time to travel right across it. A causally connected region will become uniform in temperature because photons can flow right through the region. It was this tiny piece of space, the size of a sub-atomic particle, that inflated in scale to become everything we observe today.

The energy density of the universe determines its geometry and how its geometry changes over time. Today we measure the geometry of the universe to be close to, or perhaps exactly, flat. This is called 'the flatness problem' – why did the geometry of the universe start out almost exactly flat? Inflation provides a mechanism that can give rise to a perfectly flat spacetime, no matter its previous curvature. Imagine the surface of a balloon that is not inflated but wrinkly and irregular and far from flat. When the balloon is inflated, the surface becomes smoother. Imagine blowing up the balloon until it is a trillion times larger – the surface around any point on the balloon becomes extremely smooth and almost perfectly flat. Any patch of the universe that was 'wrinkly' with a curved geometry, would be blown up to an enormous size, ending up almost, but not quite, exactly flat by the end of inflation.

Many of the scientists who developed the theory of inflation have been well recognised for their contributions, others less so, and Gliner not at all. From 2002 until 2017 I served on the scientific board of the Walter Tomalla Foundation – one of three professors who award a 100,000 Swiss Franc prize to fundamental research in gravitation and cosmology. In 2009 we awarded the Tomalla Prize to Viatcheslav Mukhanov, for his contributions to inflationary cosmology and predicting the origin of primordial fluctuations, and to Alexei Starobinsky for his contributions to inflationary cosmology and predicting the spectrum of gravitational waves that would ensue. At this time, I was not aware of the work of Gliner and regret overlooking his contribution to this fascinating story. But the story of inflation is not finished yet, the ideas were incomplete – it turned out that the theories of Sato, Guth and Kazanas had a fundamental problem – one that was solved by Andre Linde and Paul Steinhardt, which led to our current theory of inflation as well as its inevitable consequence, the multiverse.

42. Andrei Linde (1948-) & Paul Steinhardt (1952-)

"for eternal inflation and the theory of the multiverse"

Our visible universe, everything that we can possibly see, is a vast space over 90 billion light years across hosting a trillion galaxies and a trillion trillion stars. The stunning idea that all of this is just a tiny patch embedded within a near infinite sea of universes has a long history, both in religion, literature and science. From ancient Greek Stoic philosophy to Hindu cosmology and Buddhist architecture, to the first scientific argument for a multiverse that was made by Ludwig Boltzmann when discussing the entropy paradox. That we might live in a multiverse was placed on a firm theoretical footing following research on the inflationary early universe.

The Stanford University based Russian-American theoretical physicist Andrei Linde and the Princeton University based American theoretical physicist Paul Steinhardt both claim to have written the first scientific paper about the theory of the multiverse. The multiverse that is an outcome of inflationary models is very different from the parallel universes of Hugh Everett. The multiverse is the idea that patches of the universe will continue to undergo new bursts of inflation, each patch generating a new universe embedded within the existing space. The multiverse gives rise to a near infinite, or perhaps infinite, number of universes. And while some of these universes may be very similar to our own, others may be very different.

Andrei Linde carried out his PhD in theoretical particle physics at the Lebedev Physical Institute in Moscow, graduating in 1975. He describes himself as an intuitive scientist, first visualising a problem before attempting to solve it mathematically. Around five years later he came up with his theory of the universe whilst discussing the early universe with a colleague on the phone in his bathroom so as not to wake his family. After his ideas crystalised he woke up his wife and said *"It seems that I know how the universe originated!"* [102]. In 1990, Linde moved to Stanford University where he became professor of physics.

Paul Steinhardt was born in New York and his parents were lawyers. As a child he recalled his father telling him wonderful bedtime stories about scientists making great discoveries. It inspired him with the desire of discovering something new that no one ever knew before. Later, as a professor at Princeton University, he would predict a new form of matter that could exist in nature called quasicrystals. Quasicrystals were subsequently discovered existing at the University of Florence Mineralogical Collection. Until then, crystals were always thought to form from a regular lattice of atoms, whereas quasicrystals have an orderly but non-periodic structure.

Linde mentions that he discussed his ideas before he published them with Stephen Hawking during his visit to Moscow in 1981. He believes that Hawking spread his ideas in America which were picked up on by Steinhardt. However, Steinhardt mentions that he had been working on the same problem independently. After receiving a preprint from Linde in 1982 he rapidly published his own work giving Linde credit where credit was due[103].

Many of the details of inflation were expounded during a three-week workshop co-organised by Stephen Hawking, which gathered experts together in Cambridge from June 21–July 9, 1982. This led to the 'death and transfiguration' of

---

[102] Oral history interviews, Andrei Linde by Alan Lightman, American Institute of Physics 1987.

[103] Oral history interviews, Paul Steinhardt by David Zierler, American Institute of Physics 2020.

inflation, the title used in the review of the workshop[104]. Inflation 'died' since detailed calculations of the density perturbations produced during an inflationary era revealed that the quantum field associated with the Higgs particle could not drive inflation as originally thought. Therefore, in the models described in the last chapter, there was no way to reheat the universe to create radiation and matter and for Lemaitre's hot big bang to begin. The old theory of inflation would give rise to a cold and empty universe. The 'transfiguration' was due to the invention of a new quantum field by Steinhardt and Linde, called the 'inflaton' field that drove the expansion of space during the early universe. In his talk at the workshop in 1982, Steinhardt first showed how inflation could be eternal and lead to a multiverse. This idea was developed in both Steinhardt's and Linde's written contributions to the workshop proceedings.

Inflation relies on the inflaton field having a characteristic form in which its potential energy can drop to its lowest energy state called a true vacuum. Before the rapid expansion occurs, the potential energy of spacetime is stuck at an unstable false minimum state called a false vacuum. It is the change in potential energy between the false and true vacuum that drives the rapid expansion of space. Then, as the inflaton field tries to reach its minimum energy level, it undergoes a series of oscillations and the vast energy present in the vacuum is transformed into the energy of radiation and gravitating matter. Linde and Steinhardt then showed that this process never ends. Quantum fluctuations cause patches of the universe to become trapped within another false vacuum state, and in these regions, inflation continues again, and

---

[104] The inflationary universe – birth, death and transfiguration. 1982, John Barrow and Michael Turner, Nature, 298, 801-805.

again. This is often called 'eternal inflation' and it is a generic outcome of inflationary models.

Space becomes filled with bubble universes, embedded within each other – continuously inflating regions each creating a brand-new universe – the multiverse. Such embedded mini-universes continue to form forever and there is no way of knowing when the entire process began, if it did even have a beginning. On the largest scales the inflationary multiverse looks rather similar to the steady state cosmology of Hoyle. But the interior of each bubble evolves according to the equations described by the standard big bang model. As each patch of the universe begins anew, it starts with a low entropy since the expansion produces a smooth landscape of particles that can then form new cosmic structures via gravity. The number of such universes is incomprehensibly vast, perhaps infinite. In the new theory of inflation, a multiverse is a generic and inevitable consequence.

Over 10,000 research papers have been written on the topic of inflation by thousands of distinct scientists. It has become the standard paradigm for the pre-big bang phase of our early universe. But not all scientists are of this opinion. After all, inflation was designed to solve the paradoxes as to why spacetime is flat and why the microwave background temperature has nearly the same value in all directions. One could argue that the universe just began with such initial conditions. Invoking a new inflaton quantum field with specific properties could be described as an extra and arbitrary complication. With, or without inflation, we still need an initial condition and until we have a theory of quantum gravity – a theory of everything – then we cannot predict which initial condition is more likely.

That we live in a multiverse has captured many of our imaginations, scientists and the public alike. However, it has a mixed reception amongst cosmologists – many think the idea is beautiful and the multiverse inevitable, others abhor the notion. Steinhardt is one of the most vocal opponents to the idea of an inflationary multiverse that he himself developed. He argues that because the multiverse allows for every possible outcome, it has no predictive power and can bever be tested. Steinhardt states that *"Scientific ideas should be simple, explanatory, predictive. The inflationary multiverse as currently understood appears to have none of those properties"*, and argues that inflation *"explains nothing and predicts nothing"*[105].

Talking with Steinhardt always reminds me of speaking with Fred Hoyle, both scientific rebels like Freeman Dyson and Fritz Zwicky. Steinhardt is one of a minority of theoretical cosmologists who challenge the now mainstream theory of inflation and who has developed a new class of early-universe theories that replace the big bang with a 'big bounce'. His new theory envisions a smooth transition from a previous period of contraction to the current period of expansion, avoiding any need for inflation and also avoiding the infamous cosmic singularity problem associated with a big bang. A natural extension of these ideas is a never beginning and never-ending cyclic universe in which epochs of bounce, expansion, and contraction repeat at regular intervals.

In January 2017 Paul Steinhardt and two colleagues published an article in Scientific American titled 'Pop goes the universe' in which they argued that the theory of inflation cannot be tested using the scientific method. One reason that

---

[105] Interview with Paul Steinhardt by John Hogan, 2014, Scientific American blog titled 'Physicist Slams Cosmic Theory He Helped Conceive'

there are so many publications on the theory of inflation is because the form of the inflaton field can be arbitrarily varied such as to give rise to completely different outcomes. They argue that no experiment can ever be designed to confirm or disprove the theory. This prompted a response in June 2017 in the same magazine written and signed by 33 prominent scientists, including Guth, Linde, Starobinsky, Weinberg & Hawking. They rejected the criticism arguing that all the observational evidence currently supports the theory.

This is all rather fascinating and brings philosophy into the heart of the debate. I suspect that we will not see a Nobel Prize awarded for the theory of inflation or the multiverse. Moreover, with the current rules, I would fear the Nobel committee would have a difficult time in determining who should be awarded the prize. Still, I believe the theory of inflation and in particular, the multiverse, is elegant, inspiring and deserves recognition. And the multiverse provides a solution to one of the greatest mysteries in cosmology, the fine-tuning problem.

That we live in a multiverse provides a natural explanation for the apparent fine tuning of the fundamental constants of nature. The value of the strength of gravity, the magnitude of the speed of light, the mass of a proton or the charge on the electron can not be predicted from any deeper theory. They are just numbers that we measure. There are over two dozen such constants and their values determine the properties of our universe. It has been noted since the 1960s that if some of these constants had different values, even by a tiny amount, then stars, planets, molecules and life would never have appeared. Take for example the charge on the electron. If it were slightly different then the reactions that allow the formation of carbon and heavier elements in stars would not occur. In such a universe there would be no life since life could not form from hydrogen and carbon alone!

This fine tuning of our universe has often been used as an argument that the universe was so designed by some supreme being. But in the multiverse, after inflation each bubble universe can end up with a different set of fundamental constants. In some universes, there is indeed no life. In others, gravity is so strong the entire region instantly collapses. In other regions, there is practically no matter at all. In others, there are human-like creatures reading a book just like this one. In a near infinite space of possibilities, nearly all possibilities arise.

Many scientists believe that if a theory cannot be tested then it has no value and is no better than a religion. I disagree since we do not know what future research might reveal. There are many examples in science of theoretical predictions, thought to be of little value, but which later turned out to be correct. Ernst Mach criticised the concept of atoms, famously countering Ludwig Boltzmann *"have you seen one?"* String theory is another example. It attempts to unify quantum mechanics and general relativity, but it has yet to make a prediction that could be measured. Does that mean it should not be pursued? Some scientists think so. Ultimately, the universe does not care what you think, science does not care about what you believe, ultimately the pursuit of knowledge is the grandest of all human activities and this endeavour is the only way that we will uncover the mysteries of our universe.

In the far future, when all the stars in our visible universe have died, once all the black holes have evaporated, the heat death of our universe is complete. Even protons might have decayed into subatomic particles. That is a sad thought, that everything we see in our stunning universe will be gone and no life can possibly exist. But the multiverse gives the possibility of new worlds emerging, of a vast space of universes in which new life flourishes, again and again for eternity. The exponential expansion of space that is stretching the galaxies further and further apart might continue faster and faster. Our own universe could be in a false vacuum state and could undergo a new phase of inflation in the future. If each atomic sized volume of space around us could inflate into a vast new universe, just imagine the near infinite number of possible universes that could emerge from our very own. Our lives span but an infinitesimally small instant of time in the multiverse, but life just like you and me has arisen many times in the past and will arise again and again in the future.

Epilogue

We live in a unique era in the history of humankind. We have realised our insignificance as we gaze out on a trillion galaxies, each containing billions of stars. We have measured the size of our visible universe and determined its age. We know when our Sun started to shine and when it will die, that the atoms in our bodies have an incredible cosmic origin We have found that our universe is expanding; the space between the galaxies is growing, and its rate of expansion is rapidly increasing. We have measured the tiny irregularities in the very early universe, the initial conditions from which all the stars and galaxies that we see have slowly emerged over cosmic time.

It is a remarkable accomplishment of the human species that we have a good understanding of the history of our universe all the way back to less than a millionth of a second after it came into existence. At the earliest moment that our understanding of physics works, our entire visible universe and everything in it, all of its matter and energy, was squeezed into a region the size of a football. That sounds amazing, and it really is. If indeed the theory of inflation is correct, then that football was once squeezed into a volume smaller than the effective size of a proton, and our visible universe may just be a speck in the ocean of a vast sea, the multiverse.

Imagine what it must have been like 10,000 years ago given the limitations that early people had to face, an ever-present quest for survival yet some spare time to think and ponder their origins. Would you have thought about how you got here? What the twinkling stars were? How night and day arose? What causes the seasons? Of course you would have, but would you have come to the conclusion that something beyond your control was responsible? Perhaps you might even have conjured up the concept of a powerful force controlling nature and ruling your destiny. Without the knowledge that humans possess today, you may well have dreamed up some 'almighty being' conveniently responsible for everything.

Since the Enlightenment, progress in our understanding of the natural world has increased dramatically, from the macroscopic scale of the entire visible universe, to the microscopic inner world of atoms. Now and again, a single individual making bold leaps, can change the way we think about the universe. But science is built on generations of thought and discovery, of wrong paths and mistakes, of luck and serendipity. We hear the names of a few scientists again and again, but the reality is that science is built on the foundations of generations of experiments, thoughts, theories and observations.

What does the future of astrophysics and cosmology hold?

Despite all our progress there is much that we do not know and there are certainly great discoveries that should and must be made. We do not know what makes up 95 percent of our universe! The nature of dark matter, be it a new particle or a modification of spacetime and gravity, is a challenging problem to solve. We have many theories as to what dark matter could be. Only new observations or laboratory experiments can shed more light on its illusive nature. And as to the mysterious dark energy, here we lack even a single compelling theory as to its nature.

The last great theoretical insights into our universe described in the final two chapters came in the early 1980s. That's four decades ago. Since that time there have been few breakthroughs in the entire field of theoretical physics, let alone cosmology. There is so much that we do not understand yet the questions that remain are extremely complex and may take many centuries to answer. What is space? What is time? What are particles and why do they have the properties they have? Perhaps the next great breakthrough will come in understanding the final pieces of the puzzle, understanding why there was an excess of matter over antimatter. Then what about a grand unified theory that links together the strong, weak and electromagnetic forces? A theory of quantum gravity? And perhaps ultimately a mathematical theory of everything?!

For many decades the Nobel Prize Committee ignored the grand discoveries in cosmology, astrophysics and astronomy. Astronomy is the oldest of all the sciences, but only since the late 1960s have prizes been given in these areas. But grand discoveries are dwindling and I suspect fewer prizes will be awarded unless the rules are changed. This makes me wonder if the Nobel Prize is still relevant in today's research environment?

Most of the awarded Nobel Prizes were certainly deserved. However, its past record reveals many flaws and biases. From its inception until 1966 when records are available, just 92 women were nominated for the Nobel Prize in physics compared to 2,745 men. Yet I hope I have convinced you that despite the inequality imposed on women across the past centuries, there have been many more women deserving of a Nobel Prize than have been awarded one. The current Nobel Prize committees are well aware of these issues, but progress is still slow.

Today there are many more scientists than ever before, and new discoveries are often made by collaborations of hundreds or even thousands of scientists. In 1920 the average number of authors per research paper was just over one. This increased to five in 2010. We now regularly see studies with over a thousand authors spanning dozens of countries. In the year 1900 there were about three new scientific papers published each day across the entire subject area of physics. One human could indeed know all of physics research at the time. In 2020 there were 10,000 new physics publications each day, of which astrophysics and cosmology count for about one percent. That's still a hundred papers per day in my own research fields – today, one human cannot even know their own broad research area but only a small subfield of it.

With its rule of a maximum of three scientists and emphasis on confirmed discoveries, the Nobel Prize does not fit with how research is carried out today. Discoveries are rarely made by a few individuals but more by large collaborations. The emphasis on confirmed discoveries is another problem. A theory is always a theory and never becomes fact. And as we have heard, it would be extremely difficult to verify many theories, such as what came before the big bang or what comes after, despite their mathematical rigour or beauty. The Nobel Prizes have certainly been a magnificent legacy of Alfred Nobel. But perhaps it is time to rethink the rules behind the world's most famous scientific awards.

Appendix I: Misconceptions about the big bang: the tireless ant analogy to our expanding universe

The big bang describes the evolution of our universe from the time it began expanding at close to its current rate – an epoch when our universe was compressed into a volume as large as a football and the nuclei of simple elements were synthesized and the cosmic photons arose. But as we have heard, this does not describe what may have been before, nor what may come after. The name 'big bang' is an unfortunate term since it conjures up the idea of an explosion from a point in space. The reason that more distant objects appear to be moving away from us faster is because there is more space between us, and it is the space that is expanding. It is rather like stretching an elastic rope. Take one and stretch it. Notice how fast your fingers are moving apart compared to the speed of a point at the middle of the band and one of your fingers; it is twice as fast since there is twice as much space between your fingers. That is why Hubble's constant is quoted in terms of speed per unit of distance.

Now imagine an ant slowly crawling along our rubber rope that can be stretched arbitrarily in length. Let's say that the ant begins its journey, crawling at one centimetre per second trying to reach the end of a one-metre-long rope. But at the same time as the ant is walking, the rope is being stretched at one metre per second. That is, after one second the ant has walked one centimetre but the rope is now two metres long. Can the ant ever reach the end of the rope? Well, not in an ant's lifetime, but if the ant could keep walking it would reach the end in about 10^{36} years!

In this example the rope is being stretched at 100 times the speed of the ant walking, so by the time the ant reaches the end of the rope, the rope is 10^{43} metres long! That's the distance the ant has travelled. This is not just the speed of the ant multiplied by the time, that would give a distance almost 100 times too small. That's because the ant is carried along by the stretching of the rope and that's the key to why the ant will always reach the end of the rope no matter how fast the rope is being stretched.

Now imagine you, the observer, at the end point of the rope, and a galaxy at the beginning of the rope and the ant being a photon of light travelling towards you. As the universe expands the space between you and the galaxy is growing ever larger. The photon has been travelling at a constant speed but space is expanding as the photon travels. Just like the ant has travelled further than its walking speed multiplied by the time, the photon has travelled further than the speed of light multiplied by the age of the universe. And that is essentially why the visible universe, a radius of 46 billion light years, is so much larger than its age multiplied by the speed of light, 13.8 billion light years.

This analogy also helps us to understand why light from distant galaxies becomes 'redshifted'. Photons can be thought of as travelling wave packets and as space expands, so does the size of that wave. It becomes stretched, so if the photon started out in the optical wavelengths it might reach you as an infra-red wave. That's why it's helpful to build telescopes on mountain tops where distant galaxies can be studied in infra-red wavelengths that are blocked by Earth's atmosphere. For distant galaxies, we can explain the stretching of wavelengths as a consequence of the stretching of space between galaxies.

Dark energy causes space to expand faster and faster. In this case the rate at which the elastic band is stretched is not uniform but it increases with time. Our poor ant will never reach the end of the band. For this reason, the universe that we can see is becoming smaller with time. The photons from the most distant galaxies will not reach us. And even if they do, the light will have been stretched into wavelengths so long and with so low energy that we cannot detect them. In about a trillion years from now, all the galaxies beyond our own will have faded away out of site. In the far future, an astronomer would have no way of determining the history of the universe.

Using the same analogy, we can understand why there is no apparent centre to the universe. Imagine yourself standing still on the stretching rope and the rope has many galaxies along its length. Where ever you stand, you will see galaxies moving away from you in both directions. That's why there is no centre to the big bang – in whichever galaxy you may be, space is expanding away in all directions. (You could just as easily imagine a two-dimensional expanding rubber sheet or a three-dimensional rubber ball.)

This conceptual point confuses many people. There is no centre to our universe. The big bang was nothing like a giant explosion at some central point from which matter was flung apart into an existing space. The big bang did not happen at one point; it happened at every point in an existing piece of space! It is space itself that is being created or stretched in between all the matter in the universe[106]. From wherever an observer chooses to stand in the universe, they would see distant galaxies moving away from them, carried away by the expansion of space. They would also see that more distant galaxies appear to be moving away faster, as expected: something 10 times as distant has 10 times as much space between it and us, so it appears to be moving away 10 times as fast. Another analogy would be to liken the expansion with an infinitely large cherry-filled fruitcake that is expanding as it cooks in an infinitely large oven. No cherry can be identified as being at the centre of the cake, each one would observe all the others moving away from it, and more distant cherries would be seen to be moving away faster.

Is there a physical edge to all this, beyond which there is nothing? This is another common question and is more difficult to answer, since we need to define what we mean by 'edge' and 'nothing'. The furthest we can see out into the universe today is the distance that light has travelled in 13.8 billion years. That distance is about 46 billion light years. We have no way of knowing if the universe is 10 times larger or infinitely large. Our universe is indeed old and very large. It could be much larger, even infinitely large.

Another conceptual problem is the fact that it is the space between galaxies that is expanding, not the space within

---

[106] Space itself can expand arbitrarily fast, but you can't travel through your existing space faster than light. General relativity is built on the principle that no information, or physical thing such as a photon, can travel faster than the limiting speed of light.

galaxies or within our solar system. In our analogy, the cherries in the cake do not expand themselves as they are tightly connected molecules – the cake expands mainly because the dough fills with air. The reason that the space within galaxies does not expand is because the gravity of all the matter in our galaxy has overcome the expansion to form the galaxy.

While this all makes sense if we accept that it is 'space' that is expanding, it is very frustrating to me that, at a deep level, we really do not understand what space is. We do not know if time or space existed before the big bang. I cannot explain to you precisely how space is stretching and what exactly it is that is being stretched. I could tell you how the surface of a balloon stretches due to the uncoiling of rubber molecules held together by intermolecular forces. But that is nothing like empty space, which we can characterise mathematically and geometrically, yet still not compare to anything physical with which we are familiar.

Appendix II: Spacetime

We are used to thinking of space being the same in all directions, like a three-dimensional grid. This is the geometry described by Euclid of Alexandria in ancient Greece around the year 300BC in his famous book 'Elements'. Euclidian geometry describes everything you have ever drawn on a flat sheet of paper. If space is described by Euclidian geometry, then we call space 'flat'. When space is flat, the internal angles of a triangle always add up to 180 degrees and parallel lines always stay the same distance apart. But if space were curved, these rules would not hold.

As a two-dimensional analogy to a curved space, consider a patient ant which has no idea that it lives on a spherical planet. Our ant thinks that its world is flat, thus the centre and exterior of its world are irrelevant. Our ant thinks that if it continues to walk in a straight line, it will just get further from its starting point. But after patiently walking for some time, right around its spherical world, our ant is surprised to find it reaches its starting point. Until then it would have had no idea that the space in which is lives is curved. But it could have determined this using geometry since angles and distances within a curved space have different properties from those within a flat space.

In our spherical planet example of curved space, if two ants started walking together in a straight line, their paths would eventually cross or diverge rather than remaining side by side. If a clever ant drew a triangle with straight lines on the surface of its world, it would add up the angles and find a value larger than 180 degrees. The sum of those angles would be related to the curvature of its space. If we had an object of a fixed size, a so called 'standard ruler', and we could observe that ruler at different distances we could measure these angles and the curvature of space directly. It turns out we do have several ways of measuring angles and sizes of very distant structures in our universe.

Einstein showed that matter and energy curve the fabric of spacetime. That curved spacetime, in turn, then dictates how matter and energy move through it. At each instant in time, the matter and energy in the universe tells spacetime how to curve, the curved spacetime tells matter how to move, and then it does: the matter and energy moves a tiny bit and the spacetime curvature changes. And then, in the next instants of time, the same equations of general relativity tell both the matter and energy and the spacetime curvature how to evolve into the future.

We care about the geometry of space because if we can measure it then we can determine the total matter and energy content of our universe. And there is a definite prediction from theoretical models as to what the geometry should be. The theory of inflation predicts the geometry of the universe to be extremely close to flat.

Where you are right now, space is certainly not flat. Earth's gravity curves space locally otherwise you would float away. But Earth's gravitational field is relatively weak in the grand scheme of things, so the curvature is rather small. Even our

Sun only distorts space by an amount that changes the apparent position of background stars. But on the largest scale, that of the visible universe, when we measure angles and distances to far away objects, when we add up the total energy of the universe, we find that the geometry is indeed flat.

Printed in Great Britain
by Amazon